Lecture Notes in Computer Science 13867

Founding Editors

Gerhard Goos
Juris Hartmanis

Editorial Board Members

The series Lecture Notes in Computer Science (LNCS), including its subseries Lecture Notes in Artificial Intelligence (LNAI) and Lecture Notes in Bioinformatics (LNBI), has established itself as a medium for the publication of new developments in computer science and information technology research, teaching, and education.

LNCS enjoys close cooperation with the computer science R & D community, the series counts many renowned academics among its volume editors and paper authors, and collaborates with prestigious societies. Its mission is to serve this international community by providing an invaluable service, mainly focused on the publication of conference and workshop proceedings and postproceedings. LNCS commenced publication in 1973.

Gabriele Lenzini · Weizhi Meng
Editors

Security
and Trust Management

18th International Workshop, STM 2022
Copenhagen, Denmark, September 29, 2022
Proceedings

 Springer

Editors
Gabriele Lenzini 🆔
University of Luxembourg
Esch-sur-Alzette, Luxembourg

Weizhi Meng 🆔
Technical University of Denmark
Kongens Lyngby, Denmark

ISSN 0302-9743   ISSN 1611-3349 (electronic)
Lecture Notes in Computer Science
ISBN 978-3-031-29503-4   ISBN 978-3-031-29504-1 (eBook)
https://doi.org/10.1007/978-3-031-29504-1

This Springer imprint is published by the registered company Springer Nature Switzerland AG
The registered company address is: Gewerbestrasse 11, 6330 Cham, Switzerland

Preface

We are pleased to introduce the proceedings of the 18th edition of the International Workshop on Security and Trust Management (STM). In recent years, COVID-19 meant that all events were run online, but this year after the lockdown was relaxed, STM enjoyed a more convivial hybrid format. Several attendees were present at the event, which was hosted at the 27th European Symposium on Research in Computer Security (ESORICS), and held in September 2022, Copenhagen, Denmark.

A spin-off of the ERCIM working group on security and trust management, STM has always welcomed scientific contributions from private and public institutions that present innovative research both theoretical and experimental.

In particular, STM accepts innovative work on several topics of interest for security and trust, some more traditional like access control, cryptographic protocols, identity management, security metrics, and privacy to cite a few; and some others, more innovative and addressing emerging demands, for instance concerning the legal and ethical aspects in security and trust research, the economics of security and trust, the interplay between them and artificial intelligence innovative solutions, as well as the social implications for trust and security.

This volume presents the contribution of this year's edition which consists of 11 accepted papers, 7 full and 4 short. They were presented at the workshop, and were organized into four sessions: security and authentication (with works on mobile device authentication, watermarking for PRFs, risk indicators for IoT devices, and decryption for mobile devices); deep learning for security and trust (vulnerability detection via deep learning, and predictive detection of image-based malware); data analysis (differential privacy based data analysis, attacks detection in Java environments, and anomaly and intrusion detection); and finally trust and security (consent for digital services, and creation of honeywords indistinguishable by an adversary).

The selection of the works reported in this volume was based on scientific quality. Each paper was reviewed in double-blind mode by at least three qualified reviewers. Finally, we had a total submission number of 18 papers, and accepted 7 full papers with an acceptance rate of 38.9%. In addition, we accepted 4 short papers.

As a tradition, also this year STM granted a Best Paper Award to one of ERCIM's most accomplished doctoral researchers. This year's award was earned by Jonas Boehler for his work "Input Secrecy & Output Privacy: Efficient Secure Computation of Differential Privacy Mechanisms". Boehler's research presents efficient and secure design and implementation of multi-party computation protocols for distributed parties that intend to compute differential privacy statistics with high accuracy.

Before we let our readers go deep into the technical part of this volume, we would like to thank all the people who helped us in the organization of STM 2022. We thank all the authors for having supported the workshop with their submissions. We thank our excellent Program Committee members for their help in selecting such a high-quality program and the quality of articles in this volume. We thank anyone who attended and presented their work on site: it was a pleasure to meet you all in person after years of online meetings, and have a live discussion.

A special thanks must go to Pierangela Samarati, chair of the Security and Trust Management Working Group, for her inspiring support and her constant guidance in the organization of the workshop. The high reputation that this workshop holds in the international community is because of her merit and work. And, eventually, we also want to thank you, the reader, for picking up this volume. We hope that the work of the many authors herein may inspire you with new ideas to advance this exciting topic of research.

September 2022   Gabriele Lenzini
                                                                  Weizhi Meng

Organization

Program Co-chairs

Gabriele Lenzini   University of Luxembourg, Luxembourg
Weizhi Meng   Technical University of Denmark (DTU),
     Denmark

Technical Program Committee

Cristina Alcaraz   University of Malaga, Spain
Chuadhry Mujeeb Ahmed   University of Strathclyde, UK
Mauro Conti   University of Padua, Italy
Sabrina De Capitani di Vimercati   Università degli Studi di Milano, Italy
Said Daoudagh   ISTI-CNR, Pisa, Italy
Jinguang Han   Southeast University, China
Marko Hölbl   University of Maribor, Slovenia
Chenglu Jin   Centrum Wiskunde & Informatica,
     The Netherlands
Panayiotis Kotzanikolaou   University of Piraeus, Greece
Wenjuan Li   Hong Kong Polytechnic University, China
Giovanni Livraga   Università degli Studi di Milano, Italy
Bo Luo   University of Kansas, USA
Xiapu Luo   Hong Kong Polytechnic University, China
Sjouke Mauw   University of Luxembourg, Luxembourg
Fabio Martinelli   IIT-CNR, Italy
Daisuke Mashima   ADSC, Singapore
Olga Gadyatskaya   LIACS, Leiden University, The Netherlands
Dieter Gollmann   TUHH, Germany
Davy Preuveneers   KU Leuven, Belgium
Joachim Posegga   University of Passau, Germany
Qingni Shen   Peking University, China
Chunhua Su   University of Aizu, Japan
Yangguang Tian   University of Surrey, UK
Zheng Yang   Southwest University, China
Chia-Mu Yu   National Chiao Tung University, Taiwan

Subreviewers

Javaria Ahmad
Carmen Cheh
Luca Degani
Denis Donadel
Reynaldo Gil-Pons
Sohaib Kiani
Zeyan Liu
Oleksii Osliak
Luca Pajola
Henrich C. Pöhls
Stewart Sentanoe

Contents

Trust and Security

Security and Authentication

SIMple ID: QR Codes for Authentication Using Basic Mobile Phones in Developing Countries

Chris Hicks[1]($\boxtimes$), Vasilios Mavroudis[1]($\boxtimes$), and Jon Crowcroft[1,2]

[1] The Alan Turing Institute, London, UK
{c.hicks,vmavroudis,jcrowcroft}@turing.ac.uk
[2] The University of Cambridge, Cambridge, UK
jon.crowcroft@cl.cam.ac.uk

Abstract. Modern foundational electronic IDentity (eID) systems commonly rely on biometric authentication so as to reduce both their deployment costs and the need for cryptographically capable end-user devices (e.g., smartcards, smartphones). However, this exposes the users to significant security and privacy risks. We introduce SIMple ID which uses existing infrastructure, Subscriber Identity Module (SIM) cards and basic feature phones, to realise modern authentication protocols without the use of biometrics. Towards this goal, we extend the international standard for displaying images stored in SIM cards and show how this can be used to generate QR codes on even basic no-frills devices. Then, we introduce a suite of lightweight eID authentication protocols designed for on-SIM execution. Finally, we discuss SIMple ID's security, benchmark its performance and explain how it can enhance the security and privacy offered by widespread foundational eID platforms such as India's Aadhaar.

Keywords: Authentication · Identity Management · Trusted platforms

1 Introduction

More than 60 less-developed countries have launched national foundational identity programs in the last 15 years [42]. Unlike functional identity, which claims specific attributes about people such as voting entitlement or drivers' licensing, foundational identity is principally concerned with asserting the uniqueness of each person [22]. In practice, often because there is no reliable civil registry to bootstrap, the uniqueness of each resident is determined using biometric deduplication during enrollment. India's Aadhaar platform for example, which has generated more than 1.3 billion unique foundational identities [11], requests samples of all ten fingerprints, both irises and a portrait photograph during enrollment. Aadhaar and the Modular Open Source Identity Platform (MOSIP) (i.e., "Aadhaar in a box" [64]) are already being trialed and adopted in six different countries [12]. They have successfully proven the foundational identity model for development and are set to impact the lives of many millions more in the near future.

G. Lenzini and W. Meng (Eds.): STM 2022, LNCS 13867, pp. 3–23, 2023.
https://doi.org/10.1007/978-3-031-29504-1_1

Aadhaar, MOSIP and other foundational electronic IDentification (eID) platforms must provide digital authentication mechanisms that extend access to government-subsidised goods and services, as well as opportunities to build credit and reputation, to some of the poorest in society [43]. The design of authentication mechanisms is hence constrained by less-developed infrastructure, including domestic power and mobile network coverage, and limited access to expensive technologies such as smartphones. Whilst smart cards have been widely adopted for authentication in developed countries [69] the same is not true globally. Developing countries are more likely to find the additional capital, skilled labour and infrastructure required by smart cards prohibitive [5, 77]. Aadhaar does not issue a smart card and, of the approximately 72 billion transactions processed to date, over 76% were based on biometric authentication. Most of the remaining transactions were authenticated by submitting sensitive personal information, such as a name and address (i.e., "demographic authentication"). In total, less than 4% of Aadhaar transactions were authenticated using One-Time-Passwords (OTPs) sent to resident's mobile phones using Short Message Service (SMS) [14].

Contemporary foundational eID systems are not making good use of basic mobile phones, even though such devices are more common than smartphones in many less-developed countries [18]. This is mostly due to the limited capabilities of basic phones (e.g., lack of secure enclaves [10]). SIMple ID is a mobile identity solution that overcomes these limitations. Based on the internationally standardised mechanism for displaying images stored on a Subscriber Identity Module (SIM) and the cryptographic capabilities of SIM cards, SIMple ID uses QR codes to transform low-cost mobile handsets into secure authentication credentials.

Contributions

1. We propose a practical extension to the international mobile communication standards that provides a secure authentication mechanism designed for the unique sociotechnical landscape of less-developed countries.
2. We evaluate our full, open-source implementation of SIMple ID[1] which includes a Java Card applet and a KaiOS patch implementing the SIM and mobile handset parts of our protocol, respectively.
3. Beyond authentication, SIMple ID provides a robust general mechanism for establishing a QR code channel between a SIM and an in-person verifier; enabling other applications such as digital payments.

2 Preliminaries

Here we provide the prerequisites for a complete understanding of SIMple ID.

2.1 Mobile Phones in Developing Countries

SIMple ID is motivated to use basic phones for authentication because they are still widely used in a number of countries. On a global scale, basic phones

---

[1] https://github.com/alan-turing-institute/simple-id.

accounted for 32% of total mobile connections in 2020 [29]. Narrowing down to the Sub-Saharan Africa region that percentage rises to 52%. In 2020, 46% of its 1.1 billion inhabitants [17] were mobile network subscribers meaning over 250 million connected using a basic device. Moving from Africa to India, whilst mobile broadband covers 99% of the country and 77% of mobile subscribers now have access to a 4G-capable handset; 49% of adults still had not adopted a smartphone in 2020 [29].

2.2   UICC and (U)SIM Cards

Beginning with 2G Global System for Mobiles (GSM) networks, Subscriber Identity Module (SIM) cards; which are based on smart cards, provided a portable mechanism for subscriber identification and authentication [47]. Originally denoting a unified hardware and software package, starting with 3G standards the two parts were separated and renamed. SIM hardware is now termed a universal integrated circuit card (UICC) [30] and is defined by standards including ISO 7816 [53] and ETSI TS 102 221 [33]. UICCs are low-cost, and low-performance, but generally provide a certified degree of tamper-resistance [48,74]. UICCs have their own Central Processing Unit (CPU), Read-Only Memory (ROM), Electrically Erasable Programmable ROM (EEPROM) and often include a dedicated cryptographic co-processor. All UICCs support the cryptographic algorithms needed for mobile network authentication including secure random number generation, cryptographic hashing and symmetric encryption. Many UICCs also support public-key algorithms such as RSA, DSA, ECDH and ECDSA [49,78].

A UICC always run at least one Network Access Application (NAA) such as the SIM [39] and Universal SIM (USIM) applications [37] for connecting to GSM and 3G+ networks, respectively. Modern UICCs are based on the Java Card Platform (JCP) which provides a Java Card Runtime Environment (JCRE), an Application Programming Interface (API) and a Java Card Virtual Machine (JCVM) [56]. The JCRE is the operating system of a Java Card and manages the shared facilities including the communication protocols, channels, interrupts, access conditions, applications and files. All Java Card UICCs also follow the GlobalPlatform standards [2] which define an API for supporting multiple applications and managing their life-cycles independently [30]. In addition, Java Card UICCs support Over-The-Air (OTA) remote management of applications and files by the network operator [35].

UICC File System. UICCs have an elementary file system based on a hierarchical file structure that is used .of all applications and data on the Java Card [33]. As shown in Fig. 1, the root of the file structure; termed the Master File (MF), is home to a number of subdirectories called Dedicated Files (DFs), application-specific subdirectories known as Application Dedicated Files (ADFs) and leaf nodes termed Elementary Files (EFs) which contain only data. The UICC API supports creating, deleting, (de)activating, reading, updating and resizing files on the UICC programmatically at run-time using an application with appropriate privileges.

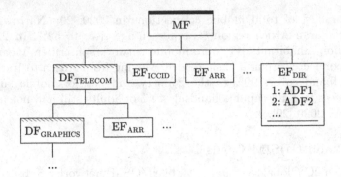

Fig. 1. The hierarchical UICC file and application structure.

Card Application Toolkit. The Card Application Toolkit (CAT) generically defines the interface between the UICC and the Mobile Equipment (ME) host (e.g., mobile phone) [34]. The CAT is commonly referred to as the SIM Toolkit (STK) [32] which more specifically designates the CAT subset supported by the GSM SIM application [39]. Similarly, the USIM Application Toolkit (USAT) [38] refers to the CAT subset supported by the 3G+ USIM application [37]. In addition to providing the UICC-ME interface which is needed for basic GSM and 3G+ network access, the CAT also provides a mechanism for the ME user to interact with the UICC through a basic menu system.

A key feature of the CAT is that it allows the UICC to proactively initiate actions taken by the ME. Since the standard smart card API is based on the model of an active host controlling a subordinate smart card [53], this is achieved through regular polling of the UICC by the ME. Proactive UICC commands for user interaction include setting up a menu, displaying text, playing tones, retrieving user input and launching the web browser [34]. Proactive UICC commands are commonly used for providing subscriber services including value-added operator content [75], mobile money [73] and digital signatures [62].

2.3   QR Codes

QR codes are an internationally standardised [46,52] form of barcode that, although originally intended to support traceability in automotive supply chains [1], have been widely adopted by mobile applications including digital identification [9,71] and payments [79]. QR codes are configured using 40 different version numbers and 4 different levels of error correcting code (ECC). Each version number prescribes a specific size, form and data capacity; meanwhile, the level of ECC determines the tolerance a code has to obscuration and physical damage. Each QR code version has a specific number of modules, ranging from 21×21 for version 1 to 177×177 for version 40, which also determine the minimum number of visual elements needed to draw the barcode. In other words a version 1 QR code requires at least 21×21 pixels on a screen, or dots on a page, for its construction. The ISO standard [52] further specifies an additional 4 modules should be left blank around the QR code on all sides to ensure readability.

2.4 Cryptographic Primitives and Notation

SIMple ID uses standardised cryptographic primitives from open standards. The use of Java Card UICCs considerably limits the scope of primitives to traditional and well-analysed algorithms [78]. The following high-level presentation aims foremost to document our notation for the familiar reader and secondly, to provide an intuition and references for the unacquainted.

Encryption schemes, used to ensure the secrecy of information, are a triple $(\text{Gen}, \text{Enc}, \text{Dec})$ of probabilistic polynomial time (PPT) algorithms for key generation, encryption and decryption, respectively. Where η is a security parameter, Gen outputs a pair of bit strings (e, d) such that $\forall (e, d) \leftarrow \text{Gen}(1^{\eta})$, and for every message $m \in \{0,1\}^{\star}$, $Pr[\text{Dec}_d(\text{Enc}_e(m)) = m] = 1$. In private key schemes $d = e$ whereas for public key schemes $d \neq e$, and e is termed the public key which we denote P_e. The private key d is denoted k_d. Informally an encryption scheme provides (semantic) security when, in the absence of k_d, encrypted messages tell an adversary nothing about the original message [45].

Digital signatures provide assurances about the authenticity of data and comprise a triple of PPT algorithms $(\text{Gen}, \text{Sign}, \text{Ver})$ for key generation, signing and verification, respectively. Gen outputs a pair of bit strings (k_s, P_v) such that $\forall (k_s, P_v) \leftarrow \text{Gen}(1^{\eta})$, and for every message $m \in \{0,1\}^{\star}$, $Pr[\text{Ver}_{P_v}(\text{Sign}_{k_s}(m)) = 1] = 1$. Informally, secure digital signature schemes demand that no adversary can forge even a single valid signature on any arbitrary message [45].

One-time passwords are a popular mechanism for authentication often encountered as a second-factor when using online password-based systems. One of the most widely used algorithms is the HMAC-Based OTP (HOTP) first described in 2005 complete with an analysis of its security [28]. In the rest of this work we use $\text{HOTP}_{k_{\text{OTP}}}(c)$ to denote the algorithm described in the standard as parameterised by private key k_{OTP} and counter c which are shared and synchronised between the client and server.

3 The Foundational eID Model

Here we introduce the authentication and adversary models of foundational eID. Aadhaar, and by extension MOSIP [67], is used as the exact basis because it is more widely described in the literature than other foundational eID systems [8, 20, 71, 72]. First we provide a brief explanation of the enrolment process before outlining the authentication and threat models that guide our development.

Regardless of the specific implementation, foundational eID is based on a Centralised IDentity Repository (CIDR) which is populated by a continuous enrollment process. Residents with identities in the CIDR are provided a Unique IDentity (UID) number that, used alongside one or more authentication factors, is used to prove identity. Enrollment is designed to ensure a one-to-one correspondence between enrolled residents and unique digital identities. Biometric enrollment, which is required to provide strong one-person-one-identity guarantees, necessitates a centralised architecture that can pairwise compare each new enrollment with the identities already in the system. In addition to collecting biometric data, including fingerprint and iris scans; enrollment also captures personal demographic information such as name, address and date of birth [6, 16].

3.1 Authentication

Authentication in the foundational paradigm requires a UID, or rarely a Virtual ID (VID), along with one or more authentication factors. The process usually takes place in-person, between a resident and a requesting entity, and a fingerprint scan is used for verification [14]. The CIDR is queried in every authentication as the scanned fingerprint biometric must be evaluated against the fingerprints associated with the submitted UID during enrollment. If the fingerprint matches, or no match is found, the CIDR returns a digitally signed "yes" or "no" response to the requesting entity, respectively. The fingerprint can be substituted with, or complemented by, the iris biometric, demographic or mobile OTP factors depending on the residents assets and the specific assurances required.

The full Aadhaar authentication ecosystem involves service agencies (SA) and user agents (UA) that form a distributed network of secure channels from the CIDR to requesting entities. In our general model of foundational authentication shown in Fig. 2, a resident R authenticates to the requesting entity RE by submitting their UID and one or more authentication factors. The RE, using a certified software and hardware stack [4], composes a digitally signed and encrypted Personal Identity Data (PID) block from the factors. Next, the *RE* sends the UID, and the encrypted PID, to the CIDR using the SA-UA network. The CIDR validates and decrypts the PID and then, using the UID to identify the correct record, checks whether the authentication factors in the PID are a match with the stored values. The CIDR digitally signs the "yes" or "no" response and returns it to the RE over the SA-UA network. Finally, the RE's authentication software validates the response and indicates the result.

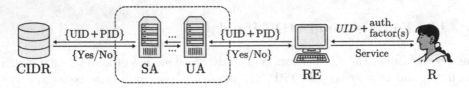

Fig. 2. The foundational eID authentication model.

3.2 Threat Model

Making no assumptions about residents access to documentation or technology means it is not practically feasible [40] to ensure uniqueness, or allow for digital authentication, without placing a lot of functionality and trust in a centralised architecture. The CIDR collects the personal and biometric information collected during enrolment and also the authentication data for every transaction. In addition, Aadhaar's SAs and UAs maintain a time-limited log of each authentication including the UID, the encrypted PID block, the CIDR's "Yes"/"No" response and any other information (e.g., e-KYC or age-verification) returned upon successful authentication [71]. REs learn all of the authentication data during every transaction except for the authentication factors encrypted in the PID. Aadhaar specifies data retention policies, access controls, biometric hardware standards and auditing practices that aim to avoid or detect vulnerabilities and inappropriate data collection. Unsurprisingly, a recent analysis identified several high-impact security and privacy breaches including subsets of the CIDR being made public and insiders making unauthorised changes to the CIDR [71]. To summarise, the foundational eID adversary model assumes the CIDR, the SA-UA network operators and the REs operate in the semi-honest model [45] by correctly executing the protocols but are able to keep a record of the intermediate computations.

4   CAT QR Codes

This section describes our technique allowing basic mobile phones to display full-screen QR codes filled with arbitrary UICC data. As described in Sect. 4.1 several proactive UICC commands support the inclusion of a graphical icon that, at the programmers discretion, can replace an otherwise text-based notification to the user. Our mechanism builds upon this standardised functionality [34], which remarkably already allows QR codes stored on the UICC as native CAT icons to be displayed using some unmodified phones, with a new image coding scheme that supports rendering image file data as a QR code.

4.1   Native Icon Protocol

A little-known feature of the CAT allows a subset of the proactive UICC commands to include an icon, such as those shown in Fig. 3, intended to enhance the user experience by providing graphical information. The UICC can also request that the icon should entirely replace the text that would otherwise be shown [34]. Icons have been supported by UICCs since GSM STK standards, albeit optionally for ME, but to the best of the authors' knowledge have received little academic or real-world attention. UICC icons support three proprietary coding schemes: black-and-white, 8 bit colour and colour-with-transparency, and are not limited in the standards to any maximum dimensions or file size [37, 39].

The standardised CAT protocol for displaying a native icon, shown in Fig. 4, involves using the DISPLAY TEXT proactive UICC command [34] that supports displaying text or an icon to the screen. To show an icon the command must include an icon qualifier bit, indicating whether the icon is "self explanatory", and an icon identifier specifying which icon to display. Self explanatory icons replace the text usually displayed by the command, potentially providing more

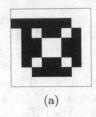  

(a)   (b)   (c)

Fig. 3. Three exemplar UICC icons from the ETSI CAT conformance specifications [36]. Icons (a) and (b) use the basic, black and white coding scheme whilst icon (c) uses the colour coding scheme and a three colour palate.

screen space for the icon. It must be noted that support for displaying icons is optional and the phone can choose to ignore the icon identifier. The terminal response may indicate to the UICC when this occurs.

Icons are selected by double reference. First, the icon identifier specifies a record number in a special lookup table called EF_{IMG}; where the width, height, coding scheme and FID of the icon is stored. Only after EF_{IMG} has been read can the icon be retrieved by the phone using the FID. Reading files from the UICC, including EF_{IMG}, is handled by the phone using ISO 7816-4 commands [53]. If the icon file is bigger than 256 bytes, the standardised maximum response payload, then it is read sequentially in chunks of 256 bytes or less. Icon files must be stored in the $DF_{GRAPHICS}$ subdirectory (see Fig. 1) with an FID 4FYY where YY is between 0 and FF.

Once the icon file has been read by the phone, the coding scheme byte specified by the corresponding EF_{IMG} record is used to render the icon to the screen. The coding scheme byte is either 0×11, 0×21 or 0×22 corresponding to black and white, colour and colour with transparency respectively. The coding schemes are all proprietary formats that begin with the width and height, optionally include colour metadata and then contain the image data using one (or several) bits per image raster point [37].

4.2 QR Code Rendering

Extending the native icon protocol to efficiently support QR codes is a relatively straightforward task that only requires defining a new image coding scheme. We do just this with a new coding scheme we term 'Render as a QR code' and assign the byte value 0x31. The protocol for rendering a QR code is unchanged from the native one described in the previous section except for the following:

1. EF_{IMG} records corresponding to QR code data must use the new coding scheme byte value.
2. Icon files no longer store image metadata and raster points but instead contain the data that will be stored in the QR code.
3. The mobile phone must implement the new coding scheme with a standard QR code rendering functionality.

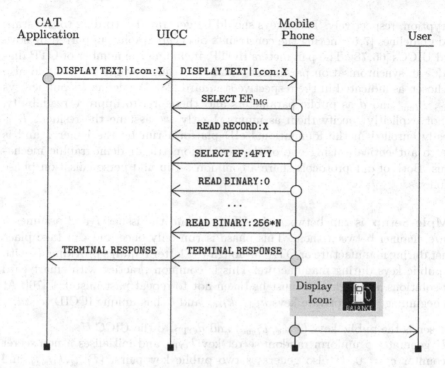

Fig. 4. The standardised CAT protocol for displaying a native icon. The application specifies the icon using the EF_{IMG} record index X that maps to the icon YY with FID 4FYY. Large icons are split into N chunks each 256 bytes or less.

Note that the changes do not deviate significantly from the existing standard. Manufacturers can thus easily implement the new icon protocol without having to allocate significant development resources. This is a significant advantage of the proposed scheme as it makes widespread adoption easier.

5 SIMple ID

Building upon our technique for displaying QR codes on basic phones, we introduce two authentication protocols in the standard foundational eID model. In particular we use the model from Sect. 3.1 but for generality, and ease of presentation, treat the SA-UA network as a secure channel. We evaluate this assumption further in Section Sect. 6.1. Both of our protocols improve upon the security and privacy offered by current foundational eID platforms. Our first protocol lightly builds upon the OTP authentication already used in Aadhaar. We move the OTP generation onto the UICC and display it along with the resident's UID in a low-version QR code. Our second protocol uses public key cryptography to provide enhanced privacy from prying requesting entities.

Firstly, our protocols assume that the CIDR issuer I has securely generated two public key pairs (P_{Isig}, k_{Isig}) and (P_{Ienc}, k_{Ienc}) for digital signatures and encryption, respectively. These keys should be generated according to the standard guidelines [7,57] noting the constraints of cryptography supported by Java Card UICCs [66,78]. The parameters HOTP, including the number of OTP digits $d \geq 6$, synchronisation parameter s and throttling parameter t, should also be chosen as indicated in the respective standard [28]. We define I's public keys $P_{I\text{sig}}$, $P_{I\text{enc}}$ and d as public parameters and therefore, to improve readability, do not explicitly specify them as inputs. Lastly, we assume the resident R is already enrolled in the foundational eID platform run by the issuer I and is able to authenticate using one of the usual biometric or demographic mechanisms. Both of our protocols share a common setup and personalisation phase as follows.

SIMple-Setup is run between the UICC U and the issuer I and assumes a secure channel between them. This phase is run only once and can take place either during manufacture or OTA, although the latter requires providing U with I's public keys during manufacture. This is common practice with smart card eID solutions although care must be taken not to repeat past mistakes [70]. At the beginning, I has private keys $k_{I\text{sig}}$, $k_{I\text{enc}}$ and U has unique ICCID $icc\text{-}id$.

1. I sends the public keys $(P_{I\text{sig}}, P_{I\text{enc}})$ and d_{OTP} to the UICC U.
2. U generates a uniform random secret key k_{OTP} and initialises a non-secret counter $c := 0$. U also generates two public key pairs, $(P_{U\text{sig}}, k_{U\text{sig}})$ and $(P_{U\text{enc}}, k_{U\text{enc}})$, for digital signatures and encryption, respectively.
3. U sends public keys $(P_{U\text{sig}}, P_{U\text{enc}})$, ICCID $icc\text{-}id$, and OTP parameters (k_{OTP}, c) to I. These values are stored as a single record by I for association with the UID of a specific resident in the next phase.

SIMple-Personalise bootstraps the resident's existing foundational authentication credential(s) to securely link their UID with the UICC U. Recall from Sect. 3.1 that the authentication credential is securely signed and encrypted, for the issuer I, using certified hardware and software provided to the requesting entity V. This phase is run between the resident R, U, V and I. Personalisation takes place after R receives the UICC and only needs to succeed once. This can be adjusted in favor of personalisation during OTP-Setup, however SIMple-Personalise allows UICCs to be quickly and widely distributed using e.g., already well-established networks of mobile agents [54]. To begin, R has UID uid and U has signing key $k_{U\text{sig}}$, encryption key $k_{U\text{enc}}$ and ICCID $icc\text{-}id$. The issuer I has signing key $k_{I\text{sig}}$, encryption key $k_{I\text{enc}}$, the CIDR containing every unique residents' UID uid and records of all $(P_{U\text{sig}}, P_{U\text{enc}})$, ICCID $icc\text{-}id$ and OTP parameters (k_{OTP}, c) submitted in the SIMple-Setup phase.

1. The resident R generates a uniformly random Personal Identification Number (PIN) pin and sends the UID uid and pin to U using their mobile phone.
2. The UICC U generates a uniformly random d-digit session identifier sid, signs the personalise message $m_{\text{pers.}} = \text{SIGN}_{k_{U\text{sig}}}(\text{``I''}|icc\text{-}id \parallel uid \parallel sid)$ and then encrypts the personalise ciphertext $c_{\text{pers.}} = \text{ENC}_{P_{I\text{enc}}}(m_{\text{pers.}})$.

3. R sends the encrypted authentication factor $c_{\text{auth.}}$ (i.e., the regular biometric or demographic PID block) and the personalise ciphertext $c_{\text{pers.}}$ (e.g., shown as a QR code using a basic phone) to the requesting entity V.

4. V sends $c_{\text{auth.}}$ and $c_{\text{pers.}}$ to the issuer I (e.g., using the SA-UA network).

5. I decrypts the personalise ciphertext to recover the personalise message $m_{\text{pers.}} = \text{DEC}_{k_{I\text{enc}}}(c_{\text{pers.}})$ containing $icc\text{-}id$, uid and sid. I uses uid to find the corresponding CIDR eID record and then verifies the regular PID $c_{\text{auth.}}$ using the standard process.

6. If this first verification succeeds (e.g., R's fingerprint matches the fingerprint(s) in the CIDR record corresponding to the uid submitted in Step 1.), then I verifies the UICC signature on $m_{\text{pers.}}$ using the public key ($P_{U\text{sig}}$ linked to $icc\text{-}id$ in the SIMple-Setup phase. Only if both of these verifications succeed will I tentatively link the UICC's $icc\text{-}id$, public keys ($P_{U\text{sig}}, P_{U\text{enc}}$) and OTP parameters ($k_{\text{OTP}}, c$) to the resident R's UID.

7. If the verifications succeeded then I signs the response message $m_{\text{resp}} = \text{SIGN}_{k_{I\text{sig}}}(\text{"yes"} \parallel sid)$ or else $m_{\text{resp}} = \text{SIGN}_{k_{I\text{sig}}}(\text{"no"})$. I sends m_{resp} to the requesting entity V.

8. V verifies the signature on m_{resp} and, if successful and m_{resp} contains the session identifier sid', sends sid' to the resident R. If unsuccessful then the protocol terminates and must begin again from Step 1.

9. R sends sid' to the UICC U using their mobile phone. If $sid' = sid$ then the resident's UID uid and PIN pin are stored by U in non-volatile memory. If unsuccessful after several attempts, the protocol terminates and must begin again from Step 1.

At this stage the UICC has been personalised for a specific resident and our two authentication protocols differ in the final phase as now described. To account for error in the final step of the SIMple-Personalise phase, for example the resident repeatedly mistypes the session id sid in Step 9, the link between the UICC and the resident is made tentatively by the issuer until the final phase has succeeded at least once. Our authentication protocols are assumed to run over a secure channel (i.e., using TLS over the standard SA-UA network).

SIMple-OTP provides a standard OTP authentication and runs between the resident R, the UICC U, the requesting entity V and the issuer I. To begin, R has the PIN pin and U has the UID uid and OTP parameters (k_{OTP}, c). I has private signing key $k_{I\text{sig}}$, the CIDR containing unique residents' UID uid and OTP parameters (k_{OTP}, c) linked in the SIMple-Personalise phase.

1. The resident R sends the PIN attempt pin' to the UICC U.

2. If $pin' \neq pin$ then authentication fails. Otherwise U computes the OTP $hotp = \text{HOTP}_{k_{\text{OTP}}}(c)$, increments the OTP counter c, and then sends the authentication message $m_{\text{auth}} = (uid \parallel hotp)$ to V.

3. V sends m_{auth} to the issuer I (e.g., using the SA-UA network).

4. I uses uid to look up the resident's eID record of the OTP parameters (k_{OTP}, c) and computes the OTP response $hotp' = \text{HOTP}_{k_{\text{OTP}}}(c)$. If the

OTP is correct, i.e., $hotp' = hotp$, then I computes the response message $m_{resp} = SIGN_{k_{Isig}}(\text{"yes"})$ and increments c. Otherwise, I computes $m_{resp} = SIGN_{k_{Isig}}(\text{"no"})$. I sends m_{resp} to the requesting entity V.

5. V verifies the signature on m_{resp} and, if successful and the message is "yes" then authentication succeeds. Otherwise authentication fails.

SIMple-VID provides authentication with improved privacy using public key encryption to hide the resident R's UID uid from the receiving entity V. This phase is run between R, the UICC U, V and the issuer I. To begin, R has the PIN pin and U has the UID uid and OTP parameters (k_{OTP}, c). I has private signing key k_{Isig}, private decryption key k_{Ienc}, the CIDR containing unique residents' UID uid and OTP parameters (k_{OTP}, c) linked in the SIMple-Personalise phase.

1. The resident R runs Step 1. from the SIMple-OTP phase.
2. If $pin' \neq pin$ then authentication fails. Otherwise U computes the OTP $hotp = HOTP_{k_{OTP}}(c)$, encrypts the authentication challenge $c_{chal} = ENC_{P_{Ienc}}(uid \parallel hotp)$ using the public key of the issuer I and increments the OTP counter c. The UICC U sends c_{chal} to the receiving entity V (i.e., it is shown as a QR code).
3. V runs Step 3. from the SIMple-OTP phase.
4. I recovers uid and $hotp$ by decrypting c_{chal} using the private encryption key k_{Ienc}. Next, I runs Step 4. from the SIMple-OTP phase.
5. V runs Step 5. from the SIMple-OTP phase.

6   Evaluation and Discussion

In this section we evaluate SIMple ID in terms of security and privacy in the foundational eID model. We also present the details of our open-source implementation, benchmark performance and discuss the use of QR codes on basic phone screens.

6.1   Security

The security of de-facto foundational eID authentication is critically dependent on the issuer fulfilling the role of a trusted third party. This is a very strong assumption, and indeed numerous insider attacks have been documented [71], however there are substantive incentives for the issuer to remain honest (but curious [45]). A common motive for less-developed countries to build a national eID platform is minimising fraud, particularly leakages in subsidy programs [58], and therefore maintaining the security of authentication is of primary importance.

It must be emphasised that the SIMple-OTP authentication protocol is simply a standard RFC 4226 HMAC-based OTP [28] accompanying the resident's (independently derived and randomly sampled) UID. A PIN on the user's UICC is used to prevent an adversary with physical device access from generating valid OTPs. The security of HOTP is formally analysed in the standard [57], which shows that no adversary without knowledge of the private key k_{OTP} can do better than approximately brute force. Where d is the number of OTP digits, s is the look-ahead synchronisation window and the adversary is allowed n total attempts, the probability of any adversary succeeding is no greater than $\frac{n*s}{10^d}$ plus some negligible advantage for exploiting the minor algorithmic bias. Whilst the PIN must be kept secret and of sufficient length, the low average transaction values of many foundational eID use cases permits a trade off with usability and convenience.

The SIMple-VID protocol builds on SIMple-OTP, to offer improved privacy from requesting entities, by encrypting the standard HOTP authentication using the public key of the issuer. From a security perspective, even if the encryption algorithm used is wholly insecure, the adversary can still do no better than to try to break the HOTP which is approximately to brute force the private key k_{OTP} as described above.

Compared with the current SMS-based OTP mechanism, security is enhanced by authenticating the UICC hardware rather than the mobile phone number it is linked with. Crucially, this avoids SIM-jacking attacks where a victim's mobile number is switched to a UICC controlled by the adversary [59]. In addition, the use of QR codes means that all 10 available OTP digits can be transmitted without impacting usability; making it around 10,000 times less likely the adversary succeeds in guessing the OTP versus when 6 digits are used.

6.2 Privacy

Firstly, the SIMple-OTP protocol does not offer any privacy benefit compared to the standard foundational eID authentication modalities. We focus therefore on the SIMple-VID protocol which essentially provides a VID-per-transaction functionality. The resident's UID is concealed from requesting entities and the SA-UA network by the use of a secure public key encryption scheme i.e., $c_{chal} = \text{ENC}_{P_{Ienc}}(uid \parallel hotp)$.

The limited cryptographic algorithms natively supported by Java Cards only includes RSA for public key encryption. Though secure when appropriate padding is used [57], RSA encryption suffers from large private key sizes relative to the security provided. The 2048 bit key size recommended for new applications [25] produces 256 byte ciphertexts, necessitating high capacity QR codes with compromised readability on basic phone screens. Fortunately the Elliptic Curve Integrated Encryption Scheme (ECIES) [65] has much smaller ciphertexts for the same level of security and has been reported to operate efficiently on the Java Card platform even without native support [41]. In any case, the lack of a standard implementation led us to prototype SIMple ID using 768 bit RSA encryption. Though insecure for use in a production environment [21], 768 bit RSA is sufficient to demonstrate the required operating principles (i.e., encrypting a digitally signed message) using nonetheless pessimistically high-capacity QR codes.

ECIES is provably secure against adaptive chosen-ciphertext attacks and provides semantic security, as-well as non-malleability [19]. Our application is straightforward and, as noted previously, offers a reduction to the security of HOTP should ECIES be fatally broken in the future.

6.3 Implementation

Our applet implementing the SIMple ID protocols was developed in Java Card and deployed on Taisys SIMoME overlay cards that comply with Java Card 3.0.4 and support the SIM toolkit (Sect. 2.2). To support our proposed extension to the icon protocol and render the QR codes (Sect. 4), we patched "TTfone TT240" devices featuring the KaiOS Operating System.

Using these implementations, we now examine whether the latency of our proposed protocols is adequately low for real-life transactions. We focus on SIMple-OTP and SIMple-VID as these are the only two protocols that will be executed repeatedly by the user once the personalization has been completed. We execute each protocol 100 times and measure the runtime of the on-SIM and the on-device parts of the protocol. As seen in Table 1, the total runtime of both protocols is less than 3 s in all cases. We observe that SIMple-VID has a longer runtime due to the extra on-SIM operations as well as the increased QR code size. Finally, we note that on-SIM execution is considerably slower than on-device operations. This is expected but highlights the importance of using the SIM only for sensitive cryptographic operations while relying on the phone's CPU for the rest of the computations.

Table 1. Average runtime (in milliseconds) of the UICC and the on-device execution components of the SIMple-OTP and SIMple-VID protocols.

Protocol	SIM Runtime (ms)	Device Runtime (ms)
SIMple-OTP	2135	220
SIMple-VID	2385	517

6.4 QR Codes

Since we display QR codes using basic phones, the screen size and pixel-density are of great importance and place limits on the maximum data capacity and readability. To establish realistic specifications, Amazon was used to identify the ten best selling basic mobile phones in India [13]. We also include the JioPhone, a handset designed to provide 4G internet access for the lowest possible cost. JioPhones are popular in India and Africa where they are heavily subsidised [55]. All of these devices have screens which are either:

- **Basic Screens** characterised by comparatively low pixels-per-inch (ppi) values between 110 and 120. The smallest screen is just 1.5 in. with a 120 × 120 resolution, although 1.77 in. and 120 × 160 is the most common.

- **Premium-Basic Screens** all have a resolution of 240 × 320, a size of 1.77 inches and a pixel-density of 167 ppi.

When considering QR codes that are displayed on such small and low-resolution screens, it is vital to ensure that each module is drawn using as many pixels as possible. Our testing using premium-basic screens found that a minimum module size of 2 × 2 pixels was required to achieve consistent readability. High levels of ECC proved to be unhelpful as the screens, even when scratched and damaged, provide high acuity compared to the industrial environments QR codes are designed to tolerate. Moreover the reduced net data capacity incurred by sacrificing modules to ECC, and the resulting need for higher version numbers and reduced module sizes, tended to worsen readability. Table 2 shows the maximum module size in pixels-squared (px^2), with an ECC level of 7%, for basic and premium-basic phone screens. Although not shown to save space, premium-basic screens can adequately render a version 13 QR code, providing a capacity of 425 bytes, with a module size of 3 px^2.

Table 2. QR code capacity, and maximum module sizes, for the two most common basic mobile phone screen resolutions.

Version number	Capacity (bytes)	N. modules	Max. module size 120 × 120 device (px^2)	Max. module size 240 × 320 device (px^2)
1	17	21 × 21	4	8
2	32	25 × 25	3	7
3	53	29 × 29	3	6
4	78	33 × 33	2	5
5	106	37 × 37	2	5
6	134	41 × 41	2	4
7	154	45 × 45	2	4
8	192	49 × 49	2	4

7   Related Work

UICCs for mobile signatures [31], also known as mobile eID [61], are already part of the national eID infrastructure in many countries. Existing solutions are primarily designed for online authentication in developed countries with users that have reliable and affordable cellular connections. Nonetheless, like SIMple ID, these systems also make use of UICCs, mobile devices and existing cellular infrastructure to provide cryptographic identity assurances [31]. Mobile-ID, a typical mobile eID currently deployed in Estonia and Azerbaijan, allows access to e-Government services and digital document signing using a basic mobile phone [60,62]. In Mobile-ID, the resident provides their mobile number to a website and then a verification code is shown on both on the website and the corresponding mobile phone. If the two codes match the resident enters their secret PIN and the UICC computes a digital signature. Finally, the digital signature is used as an authentication token for the requested website. Mobile-ID has been formally modelled and proven secure using ProVeif [62], but in contrast to SIMple ID may harm privacy by revealing a phone number to the website in every transaction. Similar UICC-based mobile eID are also deployed in many other countries including Finland [76], Moldova [63], Norway [3], Switzerland [68] and Turkey [26]. Zefferer and Teufl [80], and separately Verzeletti et al. [44], systematically review mobile eID systems. Beyond eID, UICCs are widely used for mobile money solutions such as the seminal M-PESA payment service [51]. M-PESA was transformative because it provides a way for those without bank accounts, people marginalised by conventional finance institutions, to send money digitally using only basic mobile phones. Within just two years of its 2007 launch in Kenya, 40% of adults were using the service [27]. Today M-PESA is extensively used for everyday purchases and international remittances by over 51 million customers spanning 7 different African countries [15].

In the academic literature Baqer et al. [23] describe DigiTally, an offline payment system for feature phones based on exchanging short codes between payee and recipient. The usability of DigiTally, another UICC-based CAT application like SIMple ID, is evaluated with participants at a university in Nairobi who report positively upon the usability and perceived security of the system. A notable finding of the DigiTally study, that informed the use of QR codes in this work, is that payers were observed to display their authentication codes to the recipient rather than read them verbally. In related work the authors also evaluate the security and usability tradeoffs of the short authentication codes used for DigiTally [24]. Beyond work focused on less-developed countries, Hassinen and Hyppönen present a protocol which, based on Finland's national PKI register and a Java Card application, provides authentication and non-repudiation using SMS messages [50].

8 Conclusion

SIMple ID can improve the security and privacy properties of existing foundational eID systems without requiring any additional investments in infrastructure. Instead, it employs technologies that the users are already using and are familiar with. Given the scepticism towards biometrics and the pressure for more privacy-preserving systems, we believe that SIMple ID can provide a viable alternative that can reach millions of users. Our techniques furthermore establish a generic platform for displaying QR codes using basic mobile phones that is readily extensible to support new applications such as in finance and targeting aid.

Beyond the technical challenges, a full-scale deployment will require properly incentivised device manufacturers and network operators. In particular, mobile network operators issuing an over-the-air update to their subscribers' SIM cards and handset manufacturers incorporating the icon standard updates. Nonetheless, foundational eID systems are backed by governments who can coordinate the actions required by the different parties.

Acknowledgments. This work was supported, in whole or in part, by the Bill & Melinda Gates Foundation [INV-001309]. Under the grant conditions of the Foundation, a Creative Commons Attribution 4.0 Generic License has already been assigned to the Author Accepted Manuscript version that might arise from this submission. Taisys Technologies Co. Ltd kindly donated 6 SIMoME overlay UICCs and provided technical support.

References

1. Japan Patent JP4258794A. Two-dimensional code having rectangular region provided with specific patterns to specify cell positions and distinction from background. DENSO Wave Corporation (1994)
2. GlobalPlatform Card Specification. Version 2.2.1, GlobalPlatform Inc. (2011)
3. Norwegian Mobile Bank ID: Reaching Scale through Collaboration, GSM (2014)
4. Compendium of Regulations, Circulars & Guidelines for (Authentication User Agency (AUA)/E-KYC User Agency (KUA), Authentication Service Agency (ASA) and biometric device provider) (2018). https://uidai.gov.in/images/resource/compendium_auth_19042018.pdf
5. Understanding Cost Drivers of Identification Systems (2018). https://openknowledge.worldbank.org/bitstream/handle/10986/31065/Understanding-Cost-Drivers-of-Identification-Systems.pdf
6. Aadhaar enrollment/correction/update form, Online. Government of India (2020). https://uidai.gov.in/images/aadhaar_enrolment_correction_form_version_2.1.pdf
7. Commercial National Security Algorithm (CNSA) Suite. MFS U/00/814670-15, National Security Agency (2021)
8. ID systems analysed: Aadhaar. Online, Privacy International (2021). https://privacyinternational.org/case-study/4698/id-systems-analysed-aadhaar
9. Regulation (EU) 2021/953. Official Journal of the European Union L211/1 (2021)
10. Security analysis of the KaiOS feature phone platform for DFS applications. Online, Financial Inclusion Global Initiative, Security Infrastructure and Trust Working Group (2021). https://figi.itu.int/wp-content/uploads/2021/04/Security-analysis-of-the-KaiOS-feature-phone-platform-for-DFS-applications-1.pdf
11. Aadhaar Dashboard. Online, Unique Identification Authority of India (2022a). https://uidai.gov.in/aadhaar_dashboard/index.php
12. About MOSIP, Modular Open Source Identity Platform. Online, Modular Open Source Identity Platform (2022). https://mosip.io/mosip/uploads/files/ABOUT%20MOSIP.pdf
13. Amazon.in Bestsellers: The most popular items in Basic Mobiles. Online, Amazon.in (2022). https://www.amazon.in/gp/bestsellers/electronics/1805559031
14. Daily Authentication Transaction Trend, Aadhaar Dashboard (2022). https://uidai.gov.in/aadhaar_dashboard/auth_trend.php?auth_id=dailytrend.   Note: 71,477,653,961 Total Authentication Transactions, 53,639,637,282 fingerprint-based

15. M-Pesa – Africa's leading fintech platform – marks 15 years of transforming lives. Online, Vodaphone Group (2022). https://www.vodafone.com/news/inclusion/mpesa-marks-15-years

16. MOSIP ID Object Definition. Online, Modular Open Source Identity Platform (2022). https://docs.mosip.io/1.1.5/modules/registration-processor/mosip-id-object-definition

17. Population, total - Sub-Saharan Africa. Online, World Bank (2022). https://data.worldbank.org/indicator/SP.POP.TOTL?locations=ZG. Note: 1.14 billion indicated population of Sub-Saharan Africa

18. The Mobile Economy 2022. Online, GSM Association (2022). https://www.gsma.com/mobileeconomy/wp-content/uploads/2022/02/280222-The-Mobile-Economy-2022.pdf

19. Abdalla, M., Bellare, M., Rogaway, P.: The oracle Diffie-Hellman assumptions and an analysis of DHIES. In: Naccache, D. (ed.) CT-RSA 2001. LNCS, vol. 2020, pp. 143–158. Springer, Heidelberg (2001). https://doi.org/10.1007/3-540-45353-9_12

20. Agrawal, S., Banerjee, S., Sharma, S.: Privacy and security of Aadhaar: a computer science perspective. Econ. Polit. Wkly. **52**(37), 93–102 (2017)

21. Kleinjung, T., et al.: Factorization of a 768-Bit RSA modulus. In: Rabin, T. (ed.) CRYPTO 2010. LNCS, vol. 6223, pp. 333–350. Springer, Heidelberg (2010). https://doi.org/10.1007/978-3-642-14623-7_18

22. Assisi, C., Ramnath, N.: The Aadhaar Effect: Why the World's Largest Identity Project Matters. Oxford University Press, Oxford (2018)

23. Baqer, K., Anderson, R., Mutegi, L., Payne, J.A., Sevilla, J.: DigiTally: piloting offline payments for phones. In: Thirteenth Symposium on Usable Privacy and Security (SOUPS 2017). USENIX Association (2017)

24. Baqer, K., Bezuidenhoudt, J., Anderson, R., Kuhn, M.: SMAPs: short message authentication protocols. In: Anderson, J., Matyáš, V., Christianson, B., Stajano, F. (eds.) Security Protocols 2016. LNCS, vol. 10368, pp. 119–132. Springer, Cham (2017). https://doi.org/10.1007/978-3-319-62033-6_15

25. Barker, E.: Recommendation for Key Management: Part 1 - General. NIST Special Publication 800–57 Part 1 Revision 5 (2020)

26. Birch, D.: Identity is the New Money. London Publishing Partnership (2014)

27. Camner, G., Pulver, C., Sjöblom, E.: What Makes a Successful Mobile Money Implementation? Learnings from M-PESA in Kenya and Tanzania. GSM (2013)

28. M'Raihi, D., et al.: HOTP: An HMAC-Based One-Time Password Algorithm. RFC 4226, The Internet Society (2005)

29. Delaporte, A., Bahia, K.: The State of Mobile Internet Connectivity 2021. Technical report. GSM Association (2021)

30. Edsbäcker, P.: SIM cards for cellular networks. An introduction to SIM card application development. B.Sc. thesis, Mid Sweden University (2012)

31. ETSI TR 102 203: Mobile Commerce (M-COMM); Mobile Signatures; Business and Functional Requirements. V1.1.1 (2003)

32. ETSI TS 101 476: Digital cellular telecommunications system (Phase 2+); GSM API for SIM toolkit stage 2 (3GPP TS 03.19 version 8.5.0 Release 1999) (2002)

33. ETSI TS 102 221: Smart Cards; UICC-Terminal interface; Physical and logical characteristics (Release 17). V17.1.0 (2022)

34. ETSI TS 102 223: Smart Cards; Card Application Toolkit (CAT). V15.3.0 (2019)

35. ETSI TS 102 226: Smart Cards; Remote APDU structure for UICC based applications (Release 16). V16.0.1, European Telecommunications Standards Institute (2020)

36. ETSI TS 102 384: Smart Cards; UICC-Terminal interface; Card Application Toolkit (CAT) conformance specification (Release 11). V11.0.0 (2022)
37. ETSI TS 131 102: Characteristics of the Universal Subscriber Identity Module (USIM) application (3GPP TS 31.102 version 17.5.0 Release 17) (2022)
38. ETSI TS 131 130: (U)SIM Application Programming Interface (API); (U)SIM API for Java^{TM} Card (3GPP TS 31.130 version 17.0.0 Release 17) (2022)
39. ETSI TS 151 011: Digital cellular telecommunications system (Phase 2+); Specification of the Subscriber Identity Module - Mobile Equipment (SIM-ME) interface (3GPP TS 51.011 version 4.15.0 Release 4) (2005)
40. Ford, B.: Identity and personhood in digital democracy: evaluating inclusion, equality, security, and privacy in pseudonym parties and other proofs of personhood. arXiv (2020). https://arxiv.org/abs/2011.02412
41. Gayoso Martínez, V., Hernández Encinas, L., Sánchez Ávila, C.: Java card implementation of the elliptic curve integrated encryption scheme using prime and binary finite fields. In: Herrero, Á., Corchado, E. (eds.) CISIS 2011. LNCS, vol. 6694, pp. 160–167. Springer, Heidelberg (2011). https://doi.org/10.1007/978-3-642-21323-6_20
42. Gelb, A., Metz, A.: Identification Revolution: Can Digital ID be Harnessed for Development? Brookings Institution Press, Washington (2018)
43. George, N.A., McKay, F.H.: The public distribution system and food security in India. Int. J. Environ. Res. Public Health 16(17), 3221 (2019)
44. Verzelettiet, G.M., et al.: A national mobile identity management strategy for electronic government services. In: 2018 17th IEEE International Conference on Trust, Security and Privacy in Computing And Communications (2018)
45. Goldreich, O.: Foundations of Cryptography, vol. 2. Cambridge University Press, Cambridge (2004)
46. GS1 General Specifications: The foundational GS1 standard that defines how identification keys, data attributes and barcodes must be used in business applications. Release 22.0, GS1 (2022)
47. GSM 11.11: Digital cellular telecommunications system (Phase 2+); Specification of the Subscriber Identity Module - Mobile Equipment interface;. V5.3.0 (1996)
48. Gupta, B., Quamara, M.: A taxonomy of various attacks on smart card-based applications and countermeasures. Concurr. Comput.: Pract. Experience 33(7), 1 (2021)
49. Handschuh, H., Paillier, P.: Smart card crypto-coprocessors for public-key cryptography. In: Quisquater, J.-J., Schneier, B. (eds.) CARDIS 1998. LNCS, vol. 1820, pp. 372–379. Springer, Heidelberg (2000). https://doi.org/10.1007/10721064_35
50. Hassinen, M., Hypponen, K.: Strong mobile authentication. In: 2005 2nd International Symposium on Wireless Communication Systems (2005)
51. Hughes, N., Lonie, S.: M-PESA: mobile money for the "Unbanked" turning cellphones into 24-hour tellers in Kenya. Technology, Governance, Globalization, Innovations (2007)
52. ISO/IEC 18004:2015: Information technology - Automatic identification and data capture techniques - QR Code bar code symbology specification (2015)
53. ISO/IEC 7816–4:2020: Identification cards - Integrated circuit cards - Part 4: Organization, security and commands for interchange (2020)
54. Ivatury, G., Mas, I.: The Early Experience with Branchless Banking. Focus Note No. 46, CGAP (2008)
55. James, J.: The smart feature phone revolution in developing countries: bringing the internet to the bottom of the pyramid. Inf. Soc. 36(4), 226–235 (2020)

56. Java Card Platform: Runtime Environment Specification. Version 2.2.1 (2003)
57. Kaliski, B., Staddon, J.: PKCS #1: RSA Cryptography Specifications Version 2.0. RFC 2437, The Internet Society (1998)
58. Khera, R.: Impact of Aadhaar in welfare programmes. Econ. Polit. Wkly. **52**(50), 61–70 (2017)
59. Konoth, R.K., Fischer, B., Fokkink, W., Athanasopoulos, E., Razavi, K., Bos, H.: SecurePay: strengthening two-factor authentication for arbitrary transactions. In: 2020 IEEE European Symposium on Security and Privacy (EuroS&P) (2020)
60. Krimpe, J.: Mobile ID: crucial element of m-government. In: Proceedings of the 2014 Conference on Electronic Governance and Open Society: Challenges in Eurasia. Association for Computing Machinery (2014)
61. Kubach, M., Leitold, H., Roßnagel, H., Schunck, C.H., Talamo, M.: SSEDIC.2020 on Mobile eID. In: Open Identity Summit 2015 (2015)
62. Laud, P., Roos, M.: Formal analysis of the Estonian mobile-ID protocol. In: Jøsang, A., Maseng, T., Knapskog, S.J. (eds.) NordSec 2009. LNCS, vol. 5838, pp. 271–286. Springer, Heidelberg (2009). https://doi.org/10.1007/978-3-642-04766-4_19
63. Manoil, V., Turcanu, I.: Moldova Mobile ID Case Study. World Bank (2018)
64. Martin, A.K.: Aadhaar in a box? Legitimizing digital identity in times of crisis. Surveill. Soc. **19**(1), 104–108 (2021)
65. Martínez, V.G., Álvarez, F.H., Encinas, L.H., Ávila, C.S.: A comparison of the standardized versions of ECIES (2010)
66. Mavroudis, V., Svenda, P.: JCMathLib: wrapper cryptographic library for transparent and certifiable JavaCard applets. In: 2020 IEEE European Symposium on Security and Privacy (EuroS&P) Workshops (2020)
67. MOSIP Docs 1.2.0: ID Authentication Services, Modular Open Source Identity Platform (2022). https://docs.mosip.io/1.2.0/modules/id-authentication-services
68. Murphy, A.: Swisscom Mobile ID: Enabling an Ecosystem for Secure Mobile Authentication. GSM Association (2018)
69. Naumann, I., Hogben, G.: Privacy features of European eID card specifications. Netw. Secur. **2008**(8), 9–13 (2008)
70. Parsovs, A.: Estonian electronic identity card: security flaws in key management. In: Proceedings of the 29th USENIX Conference on Security Symposium (2020)
71. Qin, K., Zhou, L., Livshits, B., Gervais, A.: India's "Aadhaar" biometric ID: structure, security, and vulnerabilities. In: Financial Cryptography and Data Security - 26th International Conference (2022)
72. Rajput, A., Gopinath, K.: Analysis of newer Aadhaar privacy models. In: Ganapathy, V., Jaeger, T., Shyamasundar, R.K. (eds.) ICISS 2018. LNCS, vol. 11281, pp. 386–404. Springer, Cham (2018). https://doi.org/10.1007/978-3-030-05171-6_20
73. Reaves, B., Scaife, N., Bates, A., Traynor, P., Butler, K.R.B.: Mo(bile) money, Mo(bile) problems: analysis of branchless banking applications in the developing world. In: 24th USENIX Security Symposium (2015)
74. Reid, J., Looi, M.: Making sense of smart card security certifications. In: Domingo-Ferrer, J., Chan, D., Watson, A. (eds.) Smart Card Research and Advanced Applications. ITIFIP, vol. 52, pp. 225–240. Springer, Boston, MA (2000). https://doi.org/10.1007/978-0-387-35528-3_13
75. Salem, A.M., Elhingary, E.A., Zerek, A.R.: Value added service for mobile communications. In: 4th International Conference on Power Engineering, Energy and Electrical Drives (2013)
76. Trichina, E., Hyppönen, K., Hassinen, M.: SIM-enabled open mobile payment system based on nation-wide PKI. In: ISSE/SECURE 2007 Securing Electronic Busi-

ness Processes, pp. 355–366. Vieweg (2007). https://doi.org/10.1007/978-3-8348-9418-2_38

77. Vashistha, A., Anderson, R., Mare, S.: Examining security and privacy research in developing regions. In: Proceedings of the 1st ACM SIGCAS Conference on Computing and Sustainable Societies. COMPASS '18 (2018)

78. Švenda, P.: Nuances of the JavaCard API on the cryptographic smart cards - JCAlgTest project. In: 7th International Workshop on Analysis of Security API (2014)

79. Wong, C.W.T., Tsui, T.C.: Automated payment over the counter - a study of Alipay, WeChat Wallet and Octopus currently used in Mainland China and Hong Kong. In: The Future of the Commercial Contract in Scholarship and Law Reform Fourth Annual Conference, Institute of Advanced Legal Studies (2019)

80. Zefferer, T., Teufl, P.: Leveraging the adoption of mobile eID and e-signature solutions in Europe. In: Kő, A., Francesconi, E. (eds.) EGOVIS 2015. LNCS, vol. 9265, pp. 86–100. Springer, Cham (2015). https://doi.org/10.1007/978-3-319-22389-6_7

A Hierarchical Watermarking Scheme for PRFs from Standard Lattice Assumptions

Yanmin Zhao[1], Yang Wang[2,3], Siu Ming Yiu[1(✉)], Yu Liu[4], and Meiqin Wang[5]

[1] Department of Computer Science, The University of Hong Kong, Hong Kong, China
{ymzhao,smyiu}@cs.hku.hk
[2] School of Mathematics, Shandong University, Jinan, China
wyang1114@sdu.edu.cn
[3] Key Laboratory of Cryptography of Zhejiang Province,
Hangzhou Normal University, Hangzhou, China
[4] School of Computer Engineering, Weifang University, Weifang, China
yuliu@wfu.edu.cn
[5] School of Cyber Science and Technology, Shandong University, Qingdao,
Shandong, China
mqwang@sdu.edu.cn

Abstract. A software watermarking scheme is to embed a "mark" or a message into a program in a cryptographic way. It is useful in proving ownership (e.g., in applications to digital rights management) and in authenticating software (e.g., for proving the distributor of the software). A qualified software watermarking scheme should satisfy three requirements: (i) the marked program should not differ from the original program significantly; (ii) the embedded "mark" or message should not be removed without destroying the program dramatically; (iii) forging a marked program without a watermarking secret key is difficult. To the best of our knowledge, existing watermarking schemes for PRFs only deal with a single key, and no scheme supports watermarking the same PRF key for multiple times which is useful for hierarchical organizations.

In the paper, we put forward a definition and security requirements for a hierarchical watermarking scheme for PRFs. Under the definition, a hierarchical watermarking scheme for PRFs is constructed to be functionality-preserving, unremovable and unforgeable under standard assumptions, namely, the LWE assumption and the SIS problem. The watermarking scheme is based on a variant translucent constrained PRF with desired security properties.

Keywords: Hierarchical Watermarking · Pseudorandom Functions · Learning with Errors (LWE) · Short Integer Solution (SIS)

1 Introduction

A software watermarking scheme is to embed a "mark" or a message into a program such that the marked program satisfies three properties: functionality-preserving, unremovability, and unforgeability. A watermarking scheme can be

G. Lenzini and W. Meng (Eds.): STM 2022, LNCS 13867, pp. 24–51, 2023.
https://doi.org/10.1007/978-3-031-29504-1_2

applied to manage digital rights, such as tracing information leaks or resolving ownership disputes. In [22], a copyrighted function which is a weaker notion than a watermarking scheme for PRFs is studied. The weakness of a copyrighted function comes from the removability. The first simulation-based software watermarking is defined by Barak et al. in [4,5]and it is stated that the *perfectly* correct software watermarking scheme does not exist even under the indistinguishability obfuscator (iO) assumption. Towards a positive result, previous works on cryptographic watermarking [23,27] present unremovable schemes by restricting an adversary's ability to modify the marked programs. Later, authors in [13] relax the watermarking definition in [4,5] by requiring that the watermarked programs should be statistically correct to the original ones. And the security model in [13] makes no restriction on adversaries. Due to the relaxation of the definition, watermarking schemes in [8,13] are proven secure without any restrictions on an adversary under the iO assumption. Since iO is not a well-studied cryptographic tool, researchers prefer to construct watermarking schemes under standard assumptions. Such results can be found in [17–19,24,26].

To the best of our knowledge, all existing watermarking schemes fail to embed another watermark into a watermarked program again. Here, we put forward a question to ask *whether there exists a watermarking scheme that enables to embed another watermark into a watermarked program again.* And this new notion of watermarking is termed as hierarchical watermarking.

Motivation. In reality, authorities that are organized hierarchically prefer a hierarchical watermarking scheme. On the other hand, a hierarchical watermarking can help delegate the workload to a subordinate watermarking center. Consider a software corporation who wants to protect their products from illegal distribution. The software corporation can embed a watermarking message (e.g. a piece of information indicating the ownership) into their products before selling them. Then, a buyer (either a community or a person) gets the product before the buyer's information watermarked into the software for a second time. It is possible that the software product can be watermarked for a third time when a community buyer allows a member to access the software. The hierarchical watermarking makes the watermarking procedure more flexible. When the software company suspects an illegal distribution, it can identify the illegal distributor by extracting the watermarked message.

Our Contributions. In the paper, we answer the proposed question positively to satisfy the practical needs of hierarchical authorities about hierarchical watermarking. We formally define a hierarchical watermarking scheme for PRFs and its security properties as well. Under our definitions, the main contribution can be summarized informally as follows:

Under the LWE assumption and the existences of PRFs and CPA-secure public-key encryptions, there is a hierarchical watermarking scheme for PRFs which is correct, unremovable and unforgeable.

Informally, the hierarchical watermarking definition captures the following intuition: a user sends a PRF key k and a message msg to the first-level watermarking center and it receives a watermarked circuit implementing the PRF functionality and other auxiliary information. After that, a user can continue to send the watermarked circuit and other messages to the subordinate levels of the watermarking center. Note that the watermarking at the first level works differently from other levels. Thus, they are defined separately. The total levels of the watermarking center is supposed to be a constant L.

The security definitions are game-based as in previous papers [17–19,24,26]. But differences exist in our security definitions. As for the unremovability, without knowing an original PRF key k, given its L watermarked circuits (unlike one circuit before) implementing the PRF under k, no adversary can remove the watermarking message from any one of the L watermarked circuits. As for the unforgeability, after seeing many pairs of a known key and its L watermarking circuits (unlike its one watermarking circuit before), no adversary can forge a watermarked circuit for a PRF key which is never queried to the watermarking center before.

Technical Overviews. The idea to construct a hierarchical watermarking scheme follows the traditional blueprint for a normal watermarking scheme. That is, the functionality of the watermarked cryptographic primitives deviates from the designed one on a negligible fraction of the whole domain. For convenience, this small fraction is rephrased as "a puncture set" which encodes the watermarking message and is kept hidden from adversaries who intend to remove the watermarking message.

Recall the watermarking scheme for PRFs proposed in [18]. The key technique is a translucent PRF which owns two types of a key: a normal key k and a constrained key ck. The normal key k computes correct function values for all points in the domain while the constrained key ck does so except for a small point set (i.e., a puncture set). Moreover, the function values evaluated at the puncture set can be tested by a testing key. Towards a watermarking scheme for translucent PRFs, the watermarking center computes two N-point sets S_0 and S_1 and the puncture set is determined by selecting the i-th point from S_{msg_i} where msg_i is the i-th bit of the watermarking message. The watermarked circuit implements a rounding of a sum of N puncture PRFs whose same key is punctured at different points in a puncture set. For more detailed, please refer to [18].

As in [18], the puncture set is one-time to watermark a program. Towards a hierarchical watermarking scheme, a naive attempt is to use the puncture set in a more subtle way. More precisely, the puncture set can be split into several subsets and each subset is treated as an individual puncture set to encode a watermarking message.

If so, then the watermarking message encoding method in [18] cannot work because an adversary can remove the watermarking message easily. Suppose an adversary has two watermarked circuits: one is watermarked at the first level of the watermarking center and the other is watermarked at the second level. Based on the attempt, since the watermarking center at the first level knows nothing about the watermarking message which will be embedded at the second level, the two puncture sets reserved for the second embedding must be different.

Then, the adversary can easily determine the puncture set used for the second embedding by just comparing the two function values and change the values once such a difference is found. In this way, the watermarking message is removed[1].

Against such an attack, a watermarking message in our scheme is encoded to be a row number of a matrix. If an adversary changes the embedded message, then it has to find a row correctly from a possible point set, which is highly impossible if the parameters are set carefully.

In detail, the watermarking center consists of L marking algorithms and L extraction algorithms. Take the marking and extraction algorithms at the first level as an example. Given a program[2] and a watermarking message, the marking algorithm at the first level computes the puncture set by computing two sets S_0, S_1 with IJL elements[3] and sorting them in $I \times J$ matrices. Each set contains L matrices which are labelled by $1, 2, \ldots, L$. Next, suppose the watermarking message is msg which is interpreted as a number. The puncture set contains all points in the first msg rows of the first matrix and the remaining $(L-1)$ matrices in S_0 and the last $(I - msg)$ rows of the first matrix in S_1. When the puncture set is determined, the program can be watermarked accordingly and the details can be found in Sect. 5.

As for the extraction, if the extraction algorithm computes the same sets S_0, S_1, then the message can be extracted correctly based on the property of the constrained key[4].

Three more questions should be considered.

- *How is the constrained key computed?* On input N different points, a constrained key is a collection of the same secret PRF key punctured at these N different points. Different from the one-time message encoding method in [18], the whole puncture set in our construction is split into L subsets. Then, at different levels, a different part of the constrained key should be output. However, in the definition of translucent PRFs in [18], the constrain key is output once for all. Hence, the definition should be adapted to allow to output constrained key partially. To this end, the constrained key generation algorithm takes as input two more arguments: one is the number set T_{in} which indicates that the secret key should be punctured at points whose positions are in T_{in} and the other is the number set T_{out} which indicates that the constrained key which is punctured at points whose positions are in T_{out} should be output.
- *How does marking algorithm executed at subordinate levels get the PRF key?* The α-th level is said to be a subordinate level of the β-th level if $\alpha > \beta$. After

---

[1] Note that the function value at a puncture point is incorrect and this incorrectness cannot be tested if the incorrect function value is modified.

[2] The program is usually determined by a pseudorandom function secret key, a signing key or a decryption key. In the paper, the program is an implementation of a PRF.

[3] Here, L is a constant and I, J, for example, can be polynomial in the security parameter.

[4] Recall that the constrained key computes incorrect function values which can be tested whether the function values are evaluated at the points in the puncture set.

watermarking at the β-th level, the marking algorithm at the subordinate α-th level is supposed to take a watermarked circuit as input. If the marking algorithm at the α-th level embeds another watermark into the watermarked circuit, then it watermarks either the circuit directly or a key. In the paper, the marking algorithm at the α-th level is formulated to watermark a key since this would be much straight forward and align with previous studies focusing on watermarking keys.

The next step is to consider how to send a key which has been watermarked at the α-th level to the $(\alpha + 1)$-th level privately. A simple solution is to use encryption schemes. Since the key also known as the plaintext is chosen by users, an encryption scheme which is secure against chosen plaintext attack (CPA) is sufficient.

- *How is the watermarking order kept?* Our scheme is designed to guarantee that if a secret PRF key k is sent to the l-th level for watermarking, then it must have been watermarked at all levels ranking higher than the l-th level. To make sure of it, the level number and the partially punctured key are encrypted together.

Organization. In Sect. 2, some preliminaries are introduced. In Sect. 3, formal definitions and security requirements for a variant translucent constrained PRF are given. In Sect. 4, the concrete construction and security analysis are stated. In Sect. 5, the formal definitions, the concrete construction and security analysis of hierarchical watermarking scheme for PRFs are given.

2   Preliminaries

Let λ be a security parameter. Let $\chi = \chi(\lambda)$ be a B-bound error distribution over integers, $n = n(\lambda)$, $m = m(\lambda)$, $q = q(\lambda)$, $p = p(\lambda)$ and $\beta = \beta(\lambda)$ be integers.

2.1   The Hypergeometric Distribution

Let N, K, n be integers. The hypergeometric distribution $\mathcal{H}(K, N, n)$ describes the number of "good" elements in n elements sampled without replacement from a set of N elements with K "good" elements. If a random variable X follows the hypergeometric distribution $\mathcal{H}(K, N, n)$, then $\Pr[X = k] = \binom{K}{k}\binom{N-K}{n-k} / \binom{N}{n}$ for $\max(0, n + K - N) \leq k \leq \min(K, n)$, where $\binom{a}{b}$ denotes a binomial coefficient for integers $b \leq a$.

Theorem 1 ([12]). *Let X be a random variable following the hypergeometric distribution $\mathcal{H}(K, N, n)$ and $\delta \geq 0$, then*

$$\Pr[X \leq \frac{K}{N} \cdot n - \delta] \leq e^{-\frac{2\delta^2}{n}}, \ \Pr[X \geq \frac{K}{N} \cdot n + \delta] \leq e^{-\frac{2\delta^2}{n}}.$$

2.2 Lattice Preliminaries

The $\text{LWE}_{n,m,q,\chi}$ assumption states that for $\mathbf{A} \xleftarrow{\$} \mathbb{Z}_q^{n \times m}$, $\mathbf{s} \xleftarrow{\$} \mathbb{Z}_q^n$, $\mathbf{e} \xleftarrow{\$} \chi^m$ and $\mathbf{u} \xleftarrow{\$} \mathbb{Z}_q^m$, the distributions for $(\mathbf{A}, \mathbf{A}^T \mathbf{s} + \mathbf{e})$ and $(\mathbf{A}, \mathbf{u})$ are computationally indistinguishable.

Suppose that $q = p\Pi_{i \in [n]} p_i$ and $p_1 < p_2 < \ldots < p_n$ are all coprime and coprime with p. The $\text{1D-SIS-R}_{m,p,q,\beta}$ problem states that given $\mathbf{v} \xleftarrow{\$} \mathbb{Z}_q^m$, compute $\mathbf{z} \in \mathbb{Z}^m$ such that $\|\mathbf{z}\| \leq \beta$ and one of the following conditions holds: (1) $\langle \mathbf{v}, \mathbf{z} \rangle \in [-\beta, \beta] + (q/p) \cdot \mathbb{Z}$; (2) $\langle \mathbf{v}, \mathbf{z} \rangle \in [-\beta, \beta] + (q/p) \cdot (\mathbb{Z} + 1/2)$. The $\text{1D-SIS-R}_{m,p,q,\beta}$ assumption states that no efficient adversary is able to solve $\text{1D-SIS-R}_{m,p,q,\beta}$ problem except with negligible probability.[5]

Theorem 2 (Lattice Trapdoors [1–3,14,20,21]). *A polynomial time algorithm TrapGen performs as follows:*

TrapGen$(1^n, 1^m, q) \to (\mathbf{W}, \mathbf{z})$: On input the parameters $n, m, q \in \mathbb{Z}$, this trapdoor generation algorithm outputs a matrix $\mathbf{W} \in \mathbb{Z}_q^{n \times m}$ and a vector $\mathbf{z} \in \mathbb{Z}^m$. Moreover,(1) the matrix $\mathbf{W}$ is statistically close to uniform;(2) the vector $\mathbf{z}$ is B-bounded, i.e., $\|\mathbf{z}\| \leq B$ and $\mathbf{W} \cdot \mathbf{z} = \mathbf{0}$ (mod q).

Theorem 3 (Homomorphic Encryption from LWE [10,15]). *The leveled homomorphic encryption scheme (HE) $\Pi_{HE} = $ (HE.KeyGen, HE.Enc, HE.Eval, HE.Dec) for (arithmetic) circuits of depth $d = d(\lambda)$ over the plaintext space $\{0,1\}^{\rho_0} \times \mathbb{Z}_q^{\rho_1}$ is defined as follows:*

- *HE.KeyGen$(1^\lambda, 1^d, 1^\rho) \to sk$: On input the security parameter λ, the circuit depth d and the length of plaintext ρ, output a secret key sk where $\rho = \rho_0 + \rho_1$.*
- *HE.Enc$(sk, (\mu, \mathbf{w})) \to ct$: On input a secret key sk and a message $(\mu, \mathbf{w}) \in \{0,1\}^{\rho_0} \times \mathbb{Z}_q^{\rho_1}$, output a ciphertext $ct \in \{0,1\}^z$ where $z = poly(\lambda, d, \rho, \log q)$.*
- *HE.Eval$(C, ct) \to ct'$: On input an arithmetic circuit $C : \{0,1\}^{\rho_0} \times \mathbb{Z}_q^{\rho_1} \to \mathbb{Z}_q$ of depth at most d, and a ciphertext $ct \leftarrow$ HE.Enc$(sk, (\mu, \mathbf{w}))$, output a ciphertext $ct' \in \{0,1\}^\tau$.*
- *HE.Dec$(sk, ct') \to w$: On input a secret key sk and a ciphertext ct', output the result which is the computation of circuit C on the plaintext $(\mu, \mathbf{w})$. Suppose the inputs to every multiplication gate in C contain at most a single non-binary value. For $sk \in \mathbb{Z}_q^\tau$, $(\mu, \mathbf{w}) \in \{0,1\}^{\rho_0} \times \mathbb{Z}_q^{\rho_1}$, compute $ct' \leftarrow$ HE.Eval$(C, $ HE.Enc$(sk, (\mu, \mathbf{w})))$. If $C(\mu, \mathbf{w}) = w \in \mathbb{Z}_q$, then with high probability, for some $E = B \cdot m^{O(d)}$,*

$$\text{HE.Dec}(sk, ct') = \langle sk, ct' \rangle = \sum_{k \in [\tau]} sk_k \cdot ct'_k \in [w - E, w + E],$$

where $\langle \cdot, \cdot \rangle$ denotes the inner product.

---

[5] In [7,11], it is proven that when $m = O(n \log q)$ and $p_1 \geq \beta \cdot \omega(\sqrt{mn \log n})$, the 1D-SIS-R$_{m,p,q,\beta}$ problem is as hard as approximating certain worst-case lattice problems to within a factor of $\beta \cdot \tilde{O}(\sqrt{mn})$.

In addition, there are two properties:

- HE.Eval$(C, \cdot)$ can be computed by a Boolean circuit of depth $poly(d, \log z)$, where z is the length of the ciphertexts output by HE.Enc.
- The scheme Π_{HE} is secure under $LWE_{n,m,q,\chi}$ assumption where $n = poly(\lambda)$ and $q > B \cdot m^{O(d)}$.

If $C : \{0,1\}^\rho \to \{0,1\}^\tau$ is a Boolean circuit, then the circuit $IP \circ C : \{0,1\}^\rho \times \mathbb{Z}_q^\tau \to \mathbb{Z}_q$ ([7,16,18]) is defined as follows:

$$(IP \circ C)(\mathbf{x}, \mathbf{y}) = IP(C(\mathbf{x}), \mathbf{y}) = \langle C(\mathbf{x}), \mathbf{y} \rangle \in \mathbb{Z}_q.$$

Theorem 4 ([6,16]). *There exist algorithms (Eval$_{pk}$, Eval$_{ct}$) such that for all matrices $\mathbf{A}_1, \ldots, \mathbf{A}_\rho, \tilde{\mathbf{A}}_1, \ldots, \tilde{\mathbf{A}}_\tau \in \mathbb{Z}_q^{n \times m}$, for all inputs $(\mathbf{x}, \mathbf{y}) \in \{0,1\}^\rho \times \mathbb{Z}_q^\tau$, and for all Boolean circuits $C : \{0,1\}^\rho \to \{0,1\}^\tau$ of depth d, if*

$$\mathbf{b}_i = \mathbf{s}^T (\mathbf{A}_i + x_i \mathbf{G}) + \mathbf{e}_i^T \ \forall i \in [\rho], \ \tilde{\mathbf{b}}_j = \mathbf{s}^T (\tilde{\mathbf{A}}_j + y_j \mathbf{G}) + \tilde{\mathbf{e}}_j^T \ \forall j \in [\tau],$$

for some vector $\mathbf{s} \in \mathbb{Z}_q^n$, and $\|\mathbf{e}_i\|, \|\tilde{\mathbf{e}}_j\| \leq B$ for all $i \in [\rho], j \in [\tau]$ where $B = B(\lambda)$ is a noise bound such that $B \cdot m^{O(d)} < q$. Define

$$\mathbf{b}_{IP \circ C} = Eval_{ct}(\mathbf{x}, IP \circ C, \mathbf{A}_1, \ldots, \mathbf{A}_\rho, \tilde{\mathbf{A}}_1, \ldots, \tilde{\mathbf{A}}_\tau, \mathbf{b}_1, \ldots, \mathbf{b}_\rho, \tilde{\mathbf{b}}_1, \ldots, \tilde{\mathbf{b}}_\tau)$$
$$\mathbf{A}_{IP \circ C} = Eval_{pk}(IP \circ C, \mathbf{A}_1, \ldots, \mathbf{A}_\rho, \tilde{\mathbf{A}}_1, \ldots, \tilde{\mathbf{A}}_\tau).$$

Then, $\mathbf{b}_{IP \circ C} = \mathbf{s}^T (\mathbf{A}_{IP \circ C} + (IP \circ C)(\mathbf{x}, \mathbf{y}) \cdot \mathbf{G}) + e_{IP \circ C}$ with $\|e_{IP \circ C}\| \leq B \cdot m^{O(d)}$.

3   A Variant of Translucent Constrained PRFs

Definition 1 (A Variant of Translucent Constrained PRFs). *A variant of translucent constrained PRF[6] with domain $\mathcal{X}$ and range $\mathcal{Y}$ is a tuple of algorithms $\Pi_{PTP} = $ (PTP.Setup, PTP.SampleKey, PTP.Eval, PTP.PCst, PTP.PCstEval, PTP.Test) with the following properties:*

- *PTP.Setup$(1^\lambda) \to (pp, tk)$: On input a security parameter λ, the setup algorithm outputs the public parameters pp and a testing key tk.*
- *PTP.SampleKey$(pp) \to msk$: On input the public parameters pp, the key sampling algorithm outputs a master PRF key msk.*
- *PTP.Eval$(pp, msk, x) \to y$: On input the public parameters pp, a master PRF key msk, and an argument $x \in \mathcal{X}$, the evaluation algorithm outputs a function value $y \in \mathcal{Y}$.*
- *PTP.PCst$(pp, msk, T, T_{in}, T_{out}) \to sk_T^{T_{out}}$: On input the public parameters pp, a master PRF key msk, and a set of points $T \subseteq \mathcal{X}$, two number sets $T_{in}, T_{out} \subseteq [L]$[7], the constraining algorithm outputs a constraint key $sk_T^{T_{out}}$.*

---

[6] The variant can generate and output partial constraint key while the original one in [18] can only generate and output the whole constraint key. The detailed discussion can be found in the introduction.

[7] T_{in} and T_{out} indicate the positions in T.

- $PTP.PCstEval(pp, sk_T, x) \rightarrow y$: *On input the public parameters pp, a constraint key sk_T, and an argument $x \in \mathcal{X}$, the constrained evaluation outputs a function value $y \in \mathcal{Y}$.*
- $PTP.Test(pp, tk, y_0) \rightarrow \{0, 1\}$: *On input the public parameters pp, a testing key tk, and a function value y_0, the testing algorithm outputs 1 for acceptance or 0 for rejection.*

In the following, the notation Π_{PTP} is used to simplify $\Pi_{PTP} = (PTP.Setup, PTP.SampleKey, PTP.Eval, PTP.PCst, PTP.PCstEval, PTP.Test)$.

Definition 2 (Selective Correctness Experiment). *Let Π_{PTP} be the variant of the translucent constrained PRF with domain $\mathcal{X}$ and range $\mathcal{Y}$. For an adversary $\mathcal{A} = (\mathcal{A}_1, \mathcal{A}_2)$ and a set system $\mathcal{S} \subseteq 2^{\mathcal{X}}$. Define the correctness experiment $Expt_{\Pi_{PTP},\mathcal{A},\mathcal{S}}$ as follows:*

Experiment $Expt_{\Pi_{PTP},\mathcal{A},\mathcal{S}}(\lambda)$:

1. $(S, st_{\mathcal{A}}) \leftarrow \mathcal{A}_1(1^\lambda)$ where $S \in \mathcal{S}$; $(pp, tk) \leftarrow PTP.Setup(1^\lambda)$; $msk \leftarrow PTP.SampleKey(pp)$;
2. $sk_S^{S_{out}} \leftarrow PTP.PCst(pp, msk, S, S_{in}, S_{out})$, and send the circuit $C_S = PTP.PCstEval(pp, sk_S^{S_{out}}, \cdot)$ to the adversary. The adversary $\mathcal{A}_2$ outputs (x, S).

Definition 3 (Correctness [18]). Π_{PTP} *is selectively correct if for any efficient adversary $\mathcal{A}$ and (x, S) output by $Expt_{\Pi_{PTP},\mathcal{A},\mathcal{S}}$:*

Evaluation Correctness:

$$\Pr[x \in \mathcal{X} \backslash S \wedge PTP.Eval(pp, msk, x) \neq PTP.PCstEval(pp, sk_S^{S_{out}}, x)] = negl(\lambda);$$

Verification Correctness:

$$\Pr[x \in S \wedge PTP.Test(pp, tk, PTP.PCstEval(pp, sk_S^{S_{out}}, x)) = 1] = 1 - negl(\lambda),$$
$$\Pr[x \in \mathcal{X} \backslash S \wedge PTP.Test(pp, tk, PTP.PCstEval(pp, sk_S^{S_{out}}, x)) = 1] = negl(\lambda).$$

3.1 Security Definitions

Definition 4 (Constrained Pseudorandom Experiment Adapted from [9,18]). *Let $b \in \{0, 1\}$ be a bit. The constrained pseudorandomness experiment $CExpt_{\Pi_{PTP},\mathcal{A},\mathcal{S}}^b(\lambda)$ is defined as follows:*

Experiment $CExpt_{\Pi_{PTP},\mathcal{A},\mathcal{S}}^b(\lambda)$:

1. $(pp, tk) \leftarrow PTP.Setup(1^\lambda)$ and $msk \leftarrow PTP.SampleKey(pp)$;
2. $(S, st_{\mathcal{A}}) \leftarrow \mathcal{A}_1^{PTP.Eval(pp,msk,\cdot)}(1^\lambda, pp)$ where $S \in \mathcal{S}$;
3. *Define the challenge oracle $\mathcal{O}_b$: (1) $\mathcal{O}_0(\cdot) = PTP.Eval(pp, msk, \cdot)$; (2) $\mathcal{O}_1(\cdot) = f(\cdot)$ where f is a truly random function. And define a circuit $Circ(\cdot) = PTP.PCstEval(pp, sk_S^{S_{out}}, \cdot)$ where $sk_S^{S_{out}} \leftarrow PTP.PCst(pp, msk, S, S_{in}, S_{out})$.*
4. *Output $b' \leftarrow \mathcal{A}^{PTP.Eval(pp,msk,\cdot),\mathcal{O}_b(\cdot)}(st_{\mathcal{A}}, Circ(\cdot))$.*

Definition 5 (Constrained Pseudorandomness Adaped from [9,18]).
The adversary is admissible if its queries to the evaluation oracle PTP.Eval(pp,
msk, ·) belong to $\mathcal{X}\backslash S$ and its queries to the challenge oracle $\mathcal{O}_0$ or $\mathcal{O}_1$ belong
to S. Then, for a fixed security parameter λ, Π_{PTP} satisfies constrained pseu-
dorandomness if for all efficient and admissible adversaries $\mathcal{A}$,

$$|\Pr[CExpt^0_{\Pi_{PTP},\mathcal{A},\mathcal{S}}(\lambda) = 1] - \Pr[CExpt^1_{\Pi_{PTP},\mathcal{A},\mathcal{S}}(\lambda) = 1]| = \mathrm{negl}(\lambda).$$

Definition 6 (Key Injectivity [18]). *Π_{PTP} is key-injective if for any two*
different keys msk_1, msk_2 sampled from the key space and x sampled from the
domain,

$$\Pr[PTP.Eval(pp, msk_1, x) = PTP.Eval(pp, msk_2, x)] = \mathrm{negl}(\lambda),$$

where the probability is taken over the randomness in PTP.Setup.

Definition 7 (Selectively Consistent Privacy Experiment Adapted
from [25]). *For an adversary $\mathcal{A}$ and a bit $b \in \{0,1\}$, define a selectively consis-*
tent privacy experiment as follows:

Experiment $CPExpt^b_{\Pi_{PTP},\mathcal{A}}(\lambda)$:

1. *To begin with, $\mathcal{A}$ submits two sets: $T_0 = \{X_{01}, X_{02}, \ldots, X_{0L}\}$ and $T_1 = \{X_{11}, X_{12}, \ldots, X_{1L}\}$ to the challenger, where each X_{ij} is also a set for $i = 0, 1$ and $j = 1, 2, \ldots, L$.*
2. *The challenger runs $(pp, tk) \leftarrow PTP.Setup(1^\lambda)$, $msk \leftarrow PTP.SampleKey(pp)$. Then, for all $j \in [L]$, the challenger computes $sk^{[L]}_{X_{bj}} \leftarrow PTP.PCst(pp, msk, X_{bj}, [L], [L])$. Finally, it sends L circuits $\{C_j\}_{j\in[L]}$ to the adversary, where $C_j(\cdot) = PTP.PCstEval(pp, sk^{[L]}_{X_{bj}}, \cdot)$.*
3. *Then, the adversary can access the evaluation oracle: when querying $x \in \{0,1\}^n$, it receives a function value $y = PTP.Eval(pp, msk, x)$.*
4. *Finally, the experiment outputs the bit b' which is the output of the adversary.*

Definition 8 (Selectively Consistent Privacy Adapted from [25]). *Define*
$d_{x,i,j} = 1$ if $x \in X_{ij}$ for $i = 0, 1$ and $j \in [L]$ and $d_{x,i,j} = 0$ otherwise. Then,
an adversary $\mathcal{A}$ is privacy-admissible if (1) for any $x \in \{0,1\}^n$, and for any
$l, j \in [L]$, $d_{x,0,j} \oplus d_{x,1,j} \oplus d_{x,0,l} \oplus d_{x,1,l} = 0$; (2) for any x submitted to the
evaluation oracle, and $j \in [L]$, $d_{x,0,j} = d_{x,1,j}$.

The Π_{PTP} is selectively consistent private if for all efficient and privacy-
admissible adversaries $\mathcal{A}$, the following holds:

$$|\Pr[CPExpt^0_{\Pi_{PTP},\mathcal{A}}(\lambda) = 1] - \Pr[CPExpt^1_{\Pi_{PTP},\mathcal{A}}(\lambda) = 1]| \le \mathrm{negl}(\lambda).$$

4 The Construction of Π_{PTP}

To start, parameters are listed as follows:

- (n, m, q, χ): LWE parameters; B_{test}: the norm bound in testing algorithm;
- ρ: length of the PRF input; p: rounding modulus;

- L: the number of hierarchical levels indexed by l; $I \times J$: the size of a puncture matrix with the elements indexed by i, j;
- N: the dimension of the coefficient vectors $\mathbf{w}$ indexed by t and $N = nm$;
- z: the bit-length of a ciphertext; τ: the bit-length of a secret key of a homomorphic encryption.

Let $\Pi_{HE} = (\mathsf{HE.KeyGen}, \mathsf{HE.Enc}, \mathsf{HE.Eval}, \mathsf{HE.Dec})$ be the (leveled) homomorphic encryption scheme which can be instantiated as in [10,15] with plaintext space $\{0,1\}^\rho \times \mathbb{Z}_q^N$. Define the equality-check circuit $\mathbf{eq}_t$ with the depth of d_{eq}: $\{0,1\}^\rho \times \{0,1\}^\rho \times \mathbb{Z}_q^N \to \mathbb{Z}_q$ where $\mathbf{eq}_t(x, (x^*, \mathbf{w})) = w_t$ if $x = x^{*8}$; 0, otherwise. Then, a homomorphic evaluation circuit C_{Eval}^t with the depth of d w.r.t. the circuit $\mathbf{eq}_t$ is defined to be $C_{Eval}^t(ct, x) = \mathsf{HE.Eval}(\mathbf{eq}_t(x, \cdot), ct)$. For $t \in [N]$, define $\mathbf{D}_t[a, b] = 1$ if $(a-1)m + b = t$; 0, otherwise.[9]

4.1   The Variant of the Translucent PRF Construction

The variant translucent PRF Π_{PTP} with domain $\{0,1\}^\rho$ and range $\mathbb{Z}_p^m$ is defined as follows:

PTP.Setup(1^λ): On input the security parameter λ, sample the following matrices uniformly at random from $\mathbb{Z}_q^{n \times m}$:

- $\hat{\mathbf{A}}$: an auxiliary matrix used to provide additional randomness;
- $\{\mathbf{A}_0, \mathbf{A}_1\}$, $\{\mathbf{B}_{i,j,r}^l\}_{i \in [I], j \in [J], r \in [z]}^{l \in [L]}$, $\{\mathbf{C}_k^l\}_{k \in [\tau]}^{l \in [L]}$: matrices for the bit encodings corresponding to the input to the PRF, ciphertexts of the punctured points under HE and the HE's secret key.

Then, take $(\mathbf{W}_{i,j}^{b,l}, \mathbf{z}_{i,j}^{b,l}) \leftarrow \mathsf{TrapGen}(1^n, 1^m, q)$ for all $b \in \{0,1\}, l \in [L], i \in [I], j \in [J]$. Finally, output the public parameters pp and a testing key tk:

$$
\begin{cases}
pp = (\hat{\mathbf{A}}, \{\mathbf{A}_0, \mathbf{A}_1\}, \{\mathbf{B}_{i,j,r}^l\}_{i \in [I], j \in [J], r \in [z]}^{l \in [L]}, \{\mathbf{C}_k^l\}_{k \in [\tau]}^{l \in [L]}, \{\mathbf{W}_{i,j}^{b,l}\}_{i \in [I], j \in [J]}^{b \in \{0,1\}, l \in [L]}) \\
tk = \{\mathbf{z}_{i,j}^{b,l}\}_{i \in [I], j \in [J]}^{b \in \{0,1\}, l \in [L]}.
\end{cases}
$$

PTP.SampleKey(pp): On input the public parameters pp, sample a PRF key $\mathbf{s} \leftarrow \chi^n$. Finally, output $msk = \mathbf{s}$.

PTP.Eval(pp, msk, x): On input the public parameters pp, the PRF key $msk = \mathbf{s}$ and an argument $x = x_1 x_2 \cdots x_\rho$, compute

$$
\tilde{\mathbf{B}}_{i,j,t}^l \leftarrow Eval_{pk}(C_t, \mathbf{B}_{i,j,1}^l, \ldots, \mathbf{B}_{i,j,z}^l, \mathbf{A}_{x_1}, \ldots, \mathbf{A}_{x_\rho}, \mathbf{C}_1^l, \ldots, \mathbf{C}_\tau^l)
$$

for all $l \in L, i \in [I], j \in [J], t \in [N]$ and $C_t = IP \circ C_{Eval}^t$. Finally, output the function value

$$
\mathbf{y}_x = \lfloor \mathbf{s}^T (\hat{\mathbf{A}} + \sum_{\substack{l \in [L], i \in [I], \\ j \in [J], t \in [N]}} \tilde{\mathbf{B}}_{i,j,t}^l \mathbf{G}^{-1}(\mathbf{D}_t)) \rceil_p.
$$

---

[8] w_t is the t-th component of the vector $\mathbf{w} \in \mathbb{Z}_q$.

[9] The collection $\{\mathbf{D}_t \in \{0,1\}^{n \times m}\}_{t \in [N]}$ is a basis for the module $\mathbb{Z}_q^{n \times m}$. Its definition makes it convenient to set a trapdoor in the function values at puncture points. More technique details can be found in [18].

PTP.PCst$(pp, msk, T, T_{in}, T_{out})$: On input the public parameters pp, the PRF key $msk = \mathbf{s}$, and puncture point set $T = \{X^{l^*}\}^{l^* \in T_{in}} = \{\ \{x_{i^*,j^*}^{b^*,l^*}\}_{i^* \in [I], j^* \in [J]}\ \}^{l^* \in T_{in}}$ where b^* is either 0 or 1, for any non-null value $x_{i^*,j^*}^{b^*,l^*}$, compute

$$\tilde{\mathbf{B}}_{i,j,i^*,j^*,t}^{b^*,l,l^*} \leftarrow Eval_{pk}(C_t, \mathbf{B}_{i,j,1}^l, \ldots, \mathbf{B}_{i,j,z}^l, \mathbf{A}_{x_{i^*,j^*,1}^{b^*,l^*}}, \ldots, \mathbf{A}_{x_{i^*,j^*,\rho}^{b^*,l^*}}, \mathbf{C}_1^l, \ldots, \mathbf{C}_\tau^l)$$

for all $l \in [L], l^* \in T_{in}, i, i^* \in [I], j, j^* \in [J], t \in [N]$ and $C_t = IP \circ C_{Eval}^t$. For any $l^* \in T_{in}, i^* \in [I], j^* \in [J]$, compute a vector $\mathbf{w}_{i^*,j^*}^{b^*,l^*} = (w_{i^*,j^*,1}^{b^*,l^*}, \ldots, w_{i^*,j^*,N}^{b^*,l^*}) \in \mathbb{Z}_q^N$ satisfying:

$$\mathbf{W}_{i^*,j^*}^{b^*,l^*} = \hat{\mathbf{A}} + \sum_{\substack{l \in [L], i \in [I] \\ j \in [J], t \in [N]}} \tilde{\mathbf{B}}_{i,j,i^*,j^*,t}^{b^*,l,l^*} \cdot G^{-1}(\mathbf{D}_t) + \sum_{t \in [N]} w_{i^*,j^*,t}^{b^*,l^*} \mathbf{D}_t.$$

Next, sample an FHE key $sk_{he} \leftarrow \mathsf{HE.KeyGen}(1^\lambda, 1^{d_{eq}}, 1^{\rho+N})$. For any $l^* \in T_{in}, i^* \in [I], j^* \in [J]$, compute $ct_{i^*,j^*}^{l^*} \leftarrow \mathsf{HE.Enc}(sk_{he}, (x_{i^*,j^*}^{b^*,l^*}, \mathbf{w}_{i^*,j^*}^{b^*,l^*}))$[10] and define $\mathbf{ct}^{l^*} = (ct_{i^*,j^*}^{l^*})_{i^* \in [I], j^* \in [J]}$.

Sample error vectors $(\mathbf{e}, \mathbf{e}_{1,0}, \mathbf{e}_{1,1}, \mathbf{e}_{2,i,j,r}^{l^*}, \mathbf{e}_{3,k}) \leftarrow \chi^m$ for any $l^* \in T_{in}, i \in [I], j \in [J], r \in [z]$ and $k \in [\tau]$. Then, compute[11]:

$$\hat{\mathbf{a}}^T = \mathbf{s}^T \hat{\mathbf{A}} + \mathbf{e}^T, \quad \mathbf{a}_b^T = \mathbf{s}^T (\mathbf{A}_b + b \cdot \mathbf{G}) + \mathbf{e}_{1,b}^T \qquad\qquad b = 0, 1$$

$$(\mathbf{b}^{l^*})_{i,j,r}^T = \mathbf{s}^T (\mathbf{B}_{i,j,r}^{l^*} + ct_{i,j,r}^{l^*} \cdot \mathbf{G}) + (\mathbf{e}^{l^*})_{2,i,j,r}^T \qquad l^* \in T_{in}, i \in [I], j \in [J], r \in [z]$$

$$\mathbf{c}_k^{l^*,T} = \mathbf{s}^T (\mathbf{C}_k^{l^*} + sk_{he,k}\mathbf{G}) + \mathbf{e}_{3,k}^T \qquad\qquad l^* \in T_{in}, k \in [\tau]$$

Next, define $\mathbf{enc}^{l^*} = (\hat{\mathbf{a}}, \{\mathbf{a}_b\}_{b \in \{0,1\}}, \{(\mathbf{b}_{i,j,r}^{l^*})_{i \in [I], j \in [J], r \in [z]}\}, \{\mathbf{c}_k^{l^*}\}_{k \in [\tau]})$ for any $l^* \in T_{in}$. Output $sk_T^{T_{out}} = \{\mathbf{enc}^{l^*}, \mathbf{ct}^{l^*}\}^{l^* \in T_{out}}$.

PTP.PCstEval(pp, sk_T, x): On input the public parameters pp, a constrained key sk_T and an argument x, compute

$$\tilde{\mathbf{b}}_{i,j,t}^l \leftarrow \mathsf{Eval}_{ct}((ct_{i,j}^l, x), C_t, \mathbf{b}_{i,j,1}^l, \ldots, \mathbf{b}_{i,j,z}^l, \mathbf{a}_{x_1}, \ldots, \mathbf{a}_{x_\rho}, \mathbf{c}_1^l, \ldots, \mathbf{c}_\tau^l)$$

for $l \in [L], i \in [I], j \in [J], t \in [N]$ and where $C_t = IP \circ C_{Eval}^t$. Finally, output the function value

$$\mathbf{y}_x = \lfloor \hat{\mathbf{a}} + \sum_{\substack{l \in [L], i \in [I] \\ j \in [J], t \in [N]}} \tilde{\mathbf{b}}_{i,j,t}^l \mathbf{G}^{-1}(\mathbf{D}_t) \rceil_p.$$

PTP.Test$(pp, tk, \mathbf{y})$: On input the testing key $tk = \{\mathbf{z}_{i,j}^{b,l}\}_{i \in [I], j \in [J]}^{b \in \{0,1\}, l \in [L]}$, a point $\mathbf{y} \in \mathbb{Z}_p^m$, output 1 if $\langle \mathbf{y}, \mathbf{z}_{i,j}^{b,l} \rangle \in [-B_{test}, B_{test}]$ for some $b \in \{0,1\}, i \in [I], j \in [J], l \in [L]$ and 0 otherwise.

Correctness and Security Analysis. The correctness and security analysis are formulated as follows.

---

[10] Note that b^* takes on either 0 or 1 and b^* is the symbol relative to the puncture point. b is the symbol standing for the bit.

[11] In the following equations, the superscript T stands for the transpostition.

Theorem 5. *Let λ be the security parameter and B be a bound on the error distribution χ. The parameter instantiations are set the same values as in [18]. Additionally, require $I, J = \omega(\log \lambda)$ and L should be some arbitrary constant. The following statements holds:*

- *Suppose $B_{test} = B(m+1), p = 2\rho^{1+\epsilon}$ for some constant $\epsilon > 0$ and $\frac{q}{2pmB} > Bm^{O(d)}$, $m' = m \cdot (3 + IJL \cdot z + L\tau)$ and $\beta = Bm^{O(d)}$, under the $LWE_{n,m',q,\chi}$ and $1D\text{-}SIS\text{-}R_{m',p,q,\beta}$ assumption, Π_{PTP} is (selectively) correct.*
- *Suppose $m' = m \cdot (3 + IJL \cdot z + \tau)$, $m'' = m \cdot (3 + IJL \cdot (z+1) + L\tau)$ and $\beta = B \cdot m^{O(d)}$. Under the the $LWE_{n,m'',q,\chi}$ and $1D\text{-}SIS\text{-}R_{m',p,q,\beta}$ assumptions, Π_{PTP} satisfies selective constrained pseudorandomness.*
- *Suppose that $\hat{p}$ is the smallest prime factor of q, B satisfies $B < \frac{\hat{p}}{2}$ and $m = \omega(n)$. Π_{PTP} is key-injective.*

Theorem 5 can be proven almost identically as in [18]. The difference between the variant and the translucent PRF in [18] is whether the constraint key is generated partially or not, which does not compromise the correctness, constrained pseudorandomness and key injectivity when the constraint key in the variant is also generated wholly.

Theorem 6 (Selectively Consistent Privacy). *If Π_{PTP} is correct and secure, then it satisfies selectively consistent privacy.*

The proof of Theorem 6 can be found in the Appendix A.

5   A Hierarchical Watermarking Scheme for PRFs

In this section, we formally define a hierarchical watermarking scheme for PRFs and its required properties.

Definition 9 (A Hierarchical Watermarking Scheme for PRFs). *For a security parameter λ, a secretly-marking, secretly-extraction, message-embedding and hierarchical watermarking scheme for PRFs is a tuple of algorithms Π_{WM} = (WM.Setup, {WM.Mark$_l$, WM.Extract$_l$}$_{l \in [L]}$) with the following properties:*

- *WM.Setup$(1^\lambda) \rightarrow \{msk^l\}_{l \in [L]}$: On input the security parameter λ, WM.Setup outputs the watermarking secret key $\{msk^l\}_{l \in [L]}$ for all levels.*
- *WM.Mark$_1(msk^1, m^1, k) \rightarrow (C^1, cipher^1)$: On input the first-level watermarking secret key msk^1, a message m^1 and the PRF key k to be marked, WM.Mark$_1$ outputs a marked circuit C^1 and a ciphertext $cipher^1$.*
- *WM.Mark$_l(msk^l, m^l, (C^{l-1}, cipher^{l-1})^{12}) \rightarrow (C^l, cipher^l)$: On input the watermarking secret key msk^l for the l-th level, a message m^l, a circuit C^{l-1} and a ciphertext $cipher^{l-1}$, WM.Mark$_l$ outputs a marked circuit C^l and a ciphertext $cipher^l$.*

---

[12] The circuit C^{l-1} and the ciphertext $cipher^{l-1}$ are output at the $(l-1)$-th level.

– $WM.Extract_l(msk^l, C', level\_number) \rightarrow \{m^l, \perp\}$: *On input the watermarking secret key msk^l, a circuit C', and a level number $level\_number$, $WM.Extract_l$ outputs the marked message m^l or a symbol $\perp$.*

In the following, the notation Π_{WM} is used to simplify $\Pi_{WM} = (WM.Setup, WM.Extract_l\}_{l \in [L]})$.

Definition 10 (Circuit Similarity [18]). *Define a circuit class $\mathcal{C} = \{C | C : \{0,1\}^n \rightarrow \{0,1\}^*\}$ and $f : \mathbb{N} \rightarrow \mathbb{N}$ is non-decreasing function. For any two circuits $C, C' \in \mathcal{C}$, the following two expressions are equivalent:*

$$C \sim_f C' \iff \Pr_{x \xleftarrow{\$} \{0,1\}^n} [C(x) \neq C'(x)] \leq 1/f(n).$$

Symmetrically, $C \nsim_f C' \iff \Pr_{x \xleftarrow{\$} \{0,1\}^n} [C(x) \neq C'(x)] \geq 1/f(n).$

Definition 11 (Watermarking Correctness). *Let Π_{WM} be a hierarchical watermarking scheme for PRFs $\Pi_{PRF} = (PRF.KeyGen, PRF.Eval)$ with domain $\{0,1\}^n$. The scheme is correct if the following two properties hold for all $l \in [L]$:*

- *$\textbf{Functionality-preserving:}$ $C^l \sim_f PRF.Eval(pp, k, \cdot)$, where C^l is the marked circuit at the l-th level, $k \leftarrow PRF.KeyGen(1^\lambda)$ and $1/f(n) = \mathrm{negl}(\lambda)$ with overwhelming probability.*
- *$\textbf{Extraction correctness:}$ $\Pr[WM.Extract_l(msk^l, C^l, l) = m^l] = 1 - \mathrm{negl}(\lambda)$, where C^l is the marked circuit and m^l is the embedded message at the l-th level.*

Security. Following [8,13,18,25], we specify two security notions for a hierarchical watermarking scheme for PRFs: unremovability and unforgeability.

Definition 12 (Watermarking Experiment). *The watermarking experiment $Expt_{\Pi_{WM}, \mathcal{A}}(\lambda)$ between an adversary $\mathcal{A}$ and a challenger $\mathcal{C}$ is defined as follows. Firstly, the challenger $\mathcal{C}$ invokes $WM.Setup(1^\lambda)$ to obtain the watermarking secret keys $\{msk^l\}_{l \in [L]}$. Then, the adversary can access:*

- *$\textbf{Marking oracle:}$ On input a message m and a secret PRF key k at the first level or on input a message m, a pair of a ciphertext and a marked circuit from the $(l-1)$-th level at the l-th level, the challenger returns a marked circuit and a ciphertext by invoking $WM.Mark_1$ or $WM.Mark_l$.*
- *$\textbf{Challenge oracle:}$ On input a set of messages $\{m^l\}_{l \in [L]}$, the challenger samples a secret key $\hat{k} \leftarrow PRF.KeyGen(1^\lambda)$ and returns watermarked circuits $\{\hat{C}^l\}_{l \in [L]}$ to the adversary.*

Finally, the adversary $\mathcal{A}$ outputs a circuit $\tilde{C}$ and a level number $\tilde{l}$. Denote by $Expt_{\Pi_{WM}, \mathcal{A}}(\lambda)$ the output of the experiment which is the output of $WM.Extract_{\tilde{l}}(msk^{\tilde{l}}, \tilde{C}, \tilde{l})$.

Definition 13 (Unremovability). *An adversary $\mathcal{A}$ is **unremoving-adm-issible** if (1) the adversary $\mathcal{A}$ makes exactly one query to the challenge oracle; (2) the circuit $\tilde{C}$ output by the adversary satisfies $\tilde{C} \sim_f \hat{C}^{\tilde{l}}$ where $\hat{C}^{\tilde{l}}$ is output by the challenger and $\tilde{l}$ is the level number output by the adversary. In addition, $1/f(n) = \text{negl}(\lambda)$.*

The hierarchical watermarking scheme Π_{WM} is unremovable if for all efficient and unremoving-admissible adversaries $\mathcal{A}$, any level number $\tilde{l} \in [L]$ output by the adversaries,

$$\Pr[\text{Expt}_{\Pi_{WM}, \mathcal{A}}(\lambda) \neq m^{\tilde{l}}] = \text{negl}(\lambda),$$

where $m^{\tilde{l}}$ is the marked message at the $\tilde{l}$-th level submitted by the adversary accessing the challenge oracle.

Definition 14 (δ-Unforgeability). *An adversary $\mathcal{A}$ is δ-**unforging-admissible** if (1) the adversary does not access the challenge oracle; (2) the circuit $\tilde{C}$ output by the adversary $\mathcal{A}$ satisfies $\tilde{C} \not\sim_f C^l_{q_l}$ for all $l \in [L], q_l \in [Q_l]$, where Q_l is the number of queries that $\mathcal{A}$ makes to the marking oracle at the l-th level, $C^l_{q_l}$ is the corresponding marked circuit for the q_l-th query at the l-th level, and $1/f > \delta$.*

The hierarchical watermarking scheme is δ-unforgeable if for all efficient and δ-unforging-admissible adversaries,

$$\Pr[\text{Expt}_{\Pi_{WM}, \mathcal{A}}(\lambda) \neq \bot] = \text{negl}(\lambda).$$

5.1 A Hierarchical Watermarking Scheme for PRFs

In this section, we demonstrate how to construct a hierarchical watermarking scheme for PRFs. The concrete construction relies on the following ingredients:

- Let Π_{PTP} be a variant translucent constrained PRF with a key space $\mathcal{K}$, a domain $\{0,1\}^n$ and a range $\{0,1\}^m$.
- Let $\Pi_{PRF} = (\text{PRF.KeyGen}, \text{PRF.Eval})$ be a secure PRF with a domain $(\{0,1\}^m)^d$ and a range $(\{0,1\}^n)^{IJ}$.
- Let $E = (\text{E.KeyGen}, \text{E.Enc}, \text{E.Dec})$ be a CPA-secure public-key encryption with a plaintext space $\mathcal{P}$ and a ciphertext space $\mathcal{C}$.

We require that $d = \frac{\lambda}{\delta}$, $\delta = \frac{1}{\text{poly}(\lambda)}$, $I = \omega(\log \lambda)$, $J = (I+1)^2 \omega(\log \lambda)$. The hierarchical watermarking scheme $\Pi_{WM} = (\text{WM.Setup}, \{\text{WM.Mark}_l, \text{WM.Extract}_l\}_{l \in [L]})$ for PRFs is defined as follows:

- WM.Setup(1^λ): On input the security parameter λ, the setup algorithm invokes $(pp, tk) \leftarrow \text{PTP.Setup}(1^\lambda)$ where $tk = \{\mathbf{z}^{b,l}_{i,j}\}^{b \in \{0,1\}, l \in [L]}_{i \in [I], j \in [J]}$. Next, for $l \in [L]$, sample L PRF keys $k^*_l \leftarrow \text{PRF.KeyGen}(1^\lambda)$ and define $K^l = \{k^*_l, k^*_{l+1}, \ldots, k^*_L\}$. For $l \in [L], u \in [d], b \in \{0,1\}$, sample $h^{b,l}_u \xleftarrow{\$} \{0,1\}^n$ uniformly at random, and define $H^{b,l} = \{h^{b,l}_u, h^{b,l+1}_u, \ldots, h^{b,L}_u\}_{u \in [d]}$ and

$TK^{b,l}=\{\mathbf{z}_{i,j}^{b,l},\mathbf{z}_{i,j}^{b,l+1},\ldots,\mathbf{z}_{i,j}^{b,L}\}_{i\in[I],j\in[J]}$. For $l=2,3,\ldots L$, sample $(pk_E^l, sk_E^l) \leftarrow$ E.KeyGen(1^λ). For $l=1$, define $(pk_E^1, sk_E^1) = (\bot, \bot)$.

Finally, it outputs the master watermarking key for every level $msk^l = (pp, \{TK^{b,l}, H^{b,l}, K^{b,l}\}_{b\in\{0,1\}}, \{pk_E^{l'}\}_{l'\in[L]}, sk_E^l)$ for all $l \in [L]$.

– WM.Mark$_1$(msk^1, m^1, k):   On   input   the key $msk^1 = (pp, \{TK^{b,1}, H^{b,1}, K^{b,1}\}_{b\in\{0,1\}}, \{pk_E^{l'}\}_{l'\in[L]}, sk_E^1)$, a PRF key $k \in \mathcal{K}$ to be marked, and a message $m^1 \in \{0,1\}^M$ where $M = \lfloor \log I \rfloor$, the marking algorithm at the first level proceeds as follows:

1. Execute the following steps:
   - for $l \in [L]$:
   - for $u \in [d]$:
   -   $y_u^{0,l} \leftarrow$ PTP.Eval($pp, k, h_u^{0,l}$)
   -   $Y^{0,l} = (y_1^{0,l}, y_2^{0,l}, \ldots, y_d^{0,l})$.

2. For $l \in [L]$, $X^{0,l} = \{x_{ij}^{0,l}\}_{i\in[I],j\in[J]} \leftarrow$ PRF.Eval($k_l^*, Y^{0,l}$).

3. Compute $y_u^{1,1} \leftarrow$ PTP.Eval($pp, k, h_u^{1,1}$) for $u \in [d]$ and define $Y^{1,1} = (y_1^{1,1}, y_2^{1,1}, \ldots, y_d^{1,1})$. Then, compute $X^{1,1} = \{x_{ij}^{1,1}\}_{i\in[I],j\in[J]} \leftarrow$ PRF.Eval $(k_1^*, Y^{1,1})$.

4. Define $T = (\{x_{ij}^{0,1}\}_{i\in[m^1],j\in[J]}, \{x_{ij}^{1,1}\}_{i\in\{m^1+1,\ldots,I\},j\in[J]}, X^{0,2}, \ldots, X^{0,L})$ and invoke PTP.PCst($pp, k, T, \{1, \ldots, L\}, \{1, \ldots, L\}$) to get the constrained key $sk_T^{[L]}$.

5. Output the circuit C^1 where $C^1(\cdot) =$ PTP.PCstEval($pp, sk_T^{[L]}, \cdot$), and a ciphertext
   $cipher^1 =$ E.Enc$_{pk_E^2}(sk_T^{\{1\}}, k, 2)$ where $sk_T^{\{1\}}$ is a part of $sk_T^{[L]}$ corresponding to the point set $(\{x_{ij}^{0,1}\}_{i\in[m^1],j\in[J]}, \{x_{ij}^{1,1}\}_{i\in\{m^1+1,\ldots,I\},j\in[J]})$.

– WM.Mark$_l$($msk^l, m^l, (C^{l-1}, cipher^{l-1})$): On input a master watermarking secret key msk^l at the l-th level, a marked circuit C^{l-1} and a cipher $cipher^{l-1}$ from the previous $(l-1)$-th level, a message m^l to be marked at the l-th level, the WM.Mark$_l$ proceeds as follows:

1. Decryption
   - Decipher the ciphertext $cipher^{l-1}$ to obtain the level number l', the secret key k and the constrained key $sk_T^{\{1\}}, \ldots, sk_T^{\{l-1\}}$. If the level number l' is not l, then output $\bot$.

2. Computation of point matrices from the l-th to the L-th level
   - for $itr = l, l+1, \ldots, L$:
   - for $u \in [d]$:
   -   $y_u^{b,itr} \leftarrow$ PTP.Eval($pp, k, h_u^{b,itr}$) for $b \in \{0,1\}$;
   -   $Y^{b,itr} = (y_1^{b,itr}, y_2^{b,itr}, \ldots, y_d^{b,itr})$ for $b \in \{0,1\}$.
   - For $itr = l, l+1, \ldots, L$, $X^{b,itr} = \{x_{ij}^{b,itr}\}_{i\in[I],j\in[J]} \leftarrow$ PRF.Eval($k_{itr}^*, Y^{b,itr}$) for $b \in \{0,1\}$.

3. Verification of puncture
   - For all $b \in \{0,1\}$, $itr = l, l+1, \ldots, L$, $i \in [I]$, initialize counters $ctr_i^{b,itr}$ and compute $ctr_i^{b,itr} = \sum_{j\in[J]}$ PTP.Test($C^{l-1}(x_{ij}^{b,itr})$). For $b=0$, if there exists some itr or i such that $ctr_i^{0,itr} \neq J$, then output $\bot$. For $b=1$, if there exists some itr or i such that $ctr_i^{1,itr} \neq 0$, then output $\bot$.

4. Computation of the constrained key
   - Define
     $T = (\{x_{ij}^{0,l}\}_{i\in[m^l],j\in[J]}, \{x_{ij}^{1,l}\}_{i\in\{m^l+1,...,I\},j\in[J]}, X^{0,l+1}, ..., X^{0,L})$ and invoke PTP.PCst$(pp, k, T, \{l, ..., L\}, \{l, ..., L\})$ to get the constrained key $sk_T^{\{l,...,L\}}$.
   - Output the circuit $C^l(\cdot) = $ PTP.PCstEval$(pp, sk_T^l, \cdot)$ where $sk_T^l = (sk_T^{\{1\}}, ..., sk_T^{\{l-1\}}, sk_T^{\{l,...,L\}})$.
5. Encryption
   - Compute $cipher^l = $ E.Enc$_{pk_E^{l+1}}(sk_T^{\{1\}}, ..., sk_T^{\{l\}}, k, l+1)$, where $sk_T^{\{l\}}$ is a part of $sk_T^{\{l,...,L\}}$ corresponding to the point set $(\{x_{ij}^{0,l}\}_{i\in[m^l],j\in[J]}, \{x_{ij}^{1,l}\}_{i\in\{m^l+1,...,I\},j\in[J]})$.
6. Finally, output the circuit $C^l(\cdot)$ and the ciphertext $cipher^l$.

- WM.Extract$_l(msk^l, C, level\_number)$: On input the master watermarking key for the l-th level and a circuit C, the extraction algorithm proceeds as follows:
1. If $level\_number < l$, then output $\perp$; otherwise,
   - for $itr = l, l+1, ..., L$:
   - for $u \in [d]$:
   - $y_u^{0,itr} \leftarrow C(h_u^{0,itr})$.
   - $Y^{0,itr} = (y_1^{0,itr}, y_2^{0,itr}, ..., y_d^{0,itr})$.
2. For $itr = l, l+1, ..., L$, $X^{0,itr} = \{x_{ij}^{0,itr}\}_{i\in[I],j\in[J]} \leftarrow$ PRF.Eval$(k_{itr}^*, Y^{0,itr})$.
3. Define $ctr_i^{itr} = \sum_{j\in[J]}$ PTP.Test$(C(x_{ij}^{0,itr}))$ for $itr = l, l+1, ..., L$ and $i \in [I]$. Specifically, set $ctr_0^{itr} = J$ and $ctr_{I+1}^{itr} = 0$.
4. Define $m^{itr} = \min_{i=0,1,...,I}\{i : |ctr_i^{itr} - ctr_{i+1}^{itr}| \geq \frac{J}{I+1}\}$. Output the embedded message m^{level_number} if $m^{level\_number} \neq 0$; otherwise, output $\perp$.

Security Analysis. The correctness and the security for the above watermarking scheme are stated in the following theorem, but the detailed proofs are deterred to Appendix B.

Theorem 7 (Correctness, Unremovability, δ-Unforgeability). *If Π_{PTP} and Π_{PRF} both are secure, and E is correct and a CPA-secure encryption scheme, then the hierarchical watermarking scheme Π_{WM} is correct, unremovable and δ-unforgeable.*

Acknowledgement. The authors are supported by the National Key R&D Program of China (No. 2021YFB3100200), the Theme-Based Research Project (T35-710/20-R), the HKU-SCF FinTech Academy, the Open Research Fund of Key Laboratory of Cryptography of Zhejiang Province (No. ZCL21010), the National Key Research and Development Program of China (No. 2021YFA1000600 and 2018YFA0704702), the National Natural Science Foundation of China (No. 61832012), the National Natural Science Foundation of China (No. 61902283) and 2019 Phd Start-up Fund of Weifang University (No. 2019BS13).

A Proof of Theorem 6

Proof. The idea behind the proof is that any two consecutive hybrids differs at one point and this difference cannot be distinguished with noticeable probability since the adversary is privacy-admissible and the security of the variant of translucent constrained PRF.

Let $\{X_{01}, X_{02}, \ldots, X_{0L}\}$ and $\{X_{11}, X_{12}, \ldots, X_{1L}\}$ be the two sets that an adversary sends to the challenger for the selectively consistent privacy experiment. Let D_j be the symmetric difference of sets X_{0j} and X_{1j}. Then, $D_j = (X_{0j} \vee X_{1j}) \setminus (X_{0j} \wedge X_{1j})$ for all $j \in [L]$ and define $D = D_1 \vee D_2 \ldots \vee D_L$.

The hybrids are defined as follows:

- **Hybrid H_0:** This is exactly the selectively consistent privacy experiment when $b = 0$. An adversary $\mathcal{A}$ chooses two sets $\{X_{01}, X_{02}, \ldots, X_{0L}\}$ and $\{X_{11}, X_{12}, \ldots, X_{1L}\}$. Then, $\mathcal{A}$ sends them to the challenger. The challenger runs $(pp, tk) \leftarrow \mathsf{PTP.Setup}(1^\lambda)$, $msk \leftarrow \mathsf{PTP.SampleKey}(pp)$. Since $b = 0$, the challenger computes $sk_{X_{0j}} \leftarrow \mathsf{PTP.PCst}(pp, msk, X_{0j})$ for all $j \in [L]$. Define a circuit $C_j(\cdot) = \mathsf{PTP.PCstEval}(pp, sk_{X_{0j}}, \cdot)$ for all $j \in [L]$ and the challenger sends all circuits $\{C_j\}_{j \in [L]}$ to the adversary. Besides, the adversary can access the evaluation oracle. Finally, the experiment outputs whatever the adversary outputs.
- **Hybrid $H_{0,i}$:** Arrange all elements in D in a lexicographical order and define D^i be the set of first i elements. Define $X_{bj}^i = (X_{bj} \vee (D_j \wedge D^i)) \setminus (X_{bj} \wedge (D_j \wedge D^i))$ for $b = 0, 1$, $j \in [L]$ and set $\{X_{01}^i, X_{02}^i, \ldots, X_{0L}^i\}$ and $\{X_{11}^i, X_{12}^i, \ldots, X_{1L}^i\}$ as the two puncture sets. The remaining experiment steps are the same as in Hybrid H_0.
- **Hybrid H_1:** This is exactly the selectively consistent privacy experiment when $b = 1$. Same as Hybrid H_0 except that the constraint key are computed as $sk_{X_{1j}} \leftarrow \mathsf{PTP.PCst}(pp, msk, X_{1j})$ for all $j \in [L]$

Observe that **Hybrid $H_{0,0}$** is the same as **Hybrid H_0** and **Hybrid $H_{0,|D|}$** is the same as **Hybrid H_1**. To see this, for any $j \in [L]$, the following equations hold:

$$\begin{cases} X_{0j}^0 = (X_{0j} \vee (D_j \wedge D^0)) \setminus (X_{0j} \wedge (D_j \wedge D^0)) = (X_{0j} \vee \emptyset) \setminus (X_{0j} \wedge \emptyset) = X_{0j}; \\ X_{0j}^{|D|} = (X_{0j} \vee (D_j \wedge D^{|D|})) \setminus (X_{0j} \wedge (D_j \wedge D^{|D|})) = (X_{0j} \vee D_j) \setminus (X_{0j} \wedge D_j) \\ = X_{1j}. \end{cases}$$

Next, we prove the indistinguishability between **Hybrid $H_{0,i}$** and **Hybrid $H_{0,i+1}$**. The difference between them is how the $(i + 1)$-th element denoted by d_{i+1} in D is computed. Since the adversary $\mathcal{A}$ is privacy-admissible, d_{i+1} must be in either $X_{0j} \wedge D_j$ or $X_{1j} \wedge D_j$ for all $j \in [L]$. In $H_{0,i}$, according to the correctness of our variant of the translucent constrained PRF,

$$C_j^i(d_{i+1}) = \begin{cases} \mathsf{PTP.PCstEval}(pp, sk_{X_{0j}^i}, d_{i+1}) & d_{i+1} \in X_{0j} \wedge D_j \\ \mathsf{PTP.Eval}(pp, msk, d_{i+1}) & d_{i+1} \in X_{1j} \wedge D_j, \end{cases}$$

where C^i_j is the j-th circuit that the challenger returns to the adversary as the challenge response in **Hybrid H$_{0,i}$**. $sk_{X^i_{0j}}$ is the constraint key for the puncture set X^i_{0j} as defined in **Hybrid H$_{0,i}$**.

In **H$_{0,i+1}$**, according to the correctness,

$$C^{i+1}_j(d_{i+1}) = \begin{cases} \text{PTP.Eval}(pp, msk, d_{i+1}) & d_{i+1} \in X_{0j} \wedge D_j \\ \text{PTP.PCstEval}(pp, sk_{X^{i+1}_{0j}}, d_{i+1}) & d_{i+1} \in X_{1j} \wedge D_j. \end{cases}$$

Define an intermediate hybrid **InterH** where, for $y_1, y_2 \xleftarrow{\$} \{0,1\}^m$,

$$C^{IH}_j(d_{i+1}) = \begin{cases} y_1 & d_{i+1} \in X_{0j} \wedge D_j \\ y_2 & d_{i+1} \in X_{1j} \wedge D_j. \end{cases}$$

Since the adversary is privacy-admissible, d_{i+1} will never be asked. Besides, since the variant of the translucent constrained PRF is constrained pseudorandom and is pseudorandom, Hybrids **H$_{0,i}$** and **H$_{0,i+1}$** are both indistinguishable with the intermediate hybrid.

B   Proof of Theorem 7

B.1   Proof of Correctness

Proof. Recall that the hierarchical watermarking scheme runs $\{msk^l\}_{l\in[L]} \leftarrow$ WM.Setup(1^λ) to get the watermarking keys. Then, a PRF key is sampled: $k \leftarrow$ PTP.SampleKey(pp). To embed a set of messages $\{m^l\}_{l\in[L]}$ to a PRF key k, invoke $\{C^l, cipher^l\}_{l\in[L]} \leftarrow$ WM.Mark$_l$ where $C^l(\cdot) = $ PTP.PCstEval$(pp, sk^l_T, \cdot)$ and sk^l_T is the constraint key at the l-th level. $\{C^l\}_{l\in[L]}$ are the watermarked circuits.

By the correctness of the encryption scheme E, the ciphertext at the l-th level can be correctly deciphered at the $(l+1)$-th level.

- **Functionality-preserving:** Let S^l be the set of points x where $C^l(x) \neq$ PTP.Eval(pp, k, x) for all $l \in [L]$ and $x \in \mathcal{D} \setminus T^l$ where $\mathcal{D}$ is the domain and T^l is the puncture point set at the l-th level. By the evaluation correctness of Π_{PTP}, it holds that $\frac{|S^l|}{2^n}$ is negligible for all $l \in [L]$. Besides, the size of T^l is at most IJL and $\frac{IJL}{2^n}$ is negligible for $I, J, L = \omega(\log \lambda)$. To sum up, $C^l(\cdot)$ agrees with PTP.Eval$(pp, k, \cdot)$ on all but a negligible fraction of points.
- **Extraction correctness:** Let X^l be the set of puncture points at l-th level and H be the set of sampled points which is part of the watermarking key used for computing X^l. Since Π_{PRF} is secure, points in X^l are pseudorandom. Moreover, points in H are sampled uniformly at random. Hence, $\Pr[x = h] \leq 2 \cdot \frac{(IJL)\cdot(Ld)}{2^n} = \text{negl}(\lambda)$ for any $x \in X^l$ and $h \in H$. By the evaluation correctness, $C^l(h) = $ PTP.Eval(pp, k, h) with high probability for $h \in H$.

Thus, with high probability, the sets of puncture points are identical in marking and extraction procedures at the same level. By the verification correctness, we get $ctr_0^l = ctr_1^l = \ldots = ctr_{m^l}^l = J$ and $ctr_{m^l+1}^l = \ldots = ctr_{l+1}^l = 0$ with high probability. To conclude, the marked message can be correctly extracted with high probability.

B.2 Proof of Unremovability

Hybrid H_0 is the watermarking experiment.

Hybrid H_1: Same as H_0, except that the challenger chooses L truly random function $\{f_l\}_{l \in [L]}$ during the setup phase. Then, during the experiment, the challenger evaluates $f_l(\cdot)$ whenever it has to evaluate $\mathsf{PRF.Eval}(k_l^*, \cdot)$.

Hybrid H_2: Same as H_1, except that for all $l \in [L]$, the challenger maintains two tables T_l^0, T_l^1 at the l-th level. Every table keeps track of a mapping $\mathcal{K} \to \{0,1\}^{nIJ}$, where $\mathcal{K}$ is the PRF key space. The challenger responds to all queries as follows:

– **Marking oracle:** Same as H_1, except that when the challenger obtains a PRF key $k \in \mathcal{K}$ either from the adversary or by decrypting a ciphertext, it firstly searches k in the tables T_l^0, T_l^1 where l is the level number from the adversary. If a match is found, then the challenger sets $X^{0,l} = T_l^0(k)$ and $X^{1,l} = T_l^1(k)$. Otherwise, the challenger uniformly samples $X^{0,l}, X^{1,l} \xleftarrow{\$} \{0,1\}^{nIJ}$, and adds the mapping $k \to X^{0,l}, k \to X^{1,l}$ to tables T_l^0, T_l^1 respectively. The rest proceeds as in H_1.

– **Challenge oracle:** On input a set of messages $\{m_l\}_{l \in [L]}$ from the adversary, the challenger samples a key $\hat{k} \leftarrow \mathsf{PTP.SampleKey}(pp)$. The puncture point set $(\hat{X}^{0,l}, \hat{X}^{1,l})$ is computed as in **Marking oracle**. The rest proceeds as in H_1.

During the extraction phase, the challenger checks whether there exist an l and two different keys from tables T_l^0, T_l^1, say, k and k', such that $Y^{0,l} = Y'^{0,l}$ or $Y^{1,l} = Y'^{1,l}$. If yes, then abort the experiment and output Bad_1. Otherwise, compute $\tilde{Y}^{0,l}, \tilde{Y}^{1,l}$ for all $l \in [L]$ as in H_1. Next, the challenger checks whether $(\tilde{Y}^{0,l}, \tilde{Y}^{1,l})$ equals some $(Y^{0,l}, Y^{1,l})$ in the table T_l^0, T_l^1 for all $l \in [L]$. If so, then set $(\tilde{X}^{0,l}, \tilde{X}^{1,l})$ to be the value $(X^{0,l}, X^{1,l})$ corresponding to the $(Y^{0,l}, Y^{1,l})$. Otherwise, uniformly sample $\tilde{X}^{0,l}, \tilde{X}^{1,l} \xleftarrow{\$} \{0,1\}^{nIJ}$. The rest of the extraction procedure is the same as H_1.

Hybrid H_3: Same as H_2, except that when answering the challenge oracle, the challenger directly samples $\{\hat{X}^{0,l}, \hat{X}^{1,l}\}_{l \in [L]} \xleftarrow{\$} \{0,1\}^{nIJ}$ without checking whether the PRF key $\hat{k}$ sampled by the challenger is queried by the adversary before. Besides, the mapping $\hat{k} \to \{\hat{X}^{0,l}, \hat{X}^{1,l}\}_{l \in [L]}$ is added into the corresponding table T_l^0, T_l^1 for $l \in [L]$ in the extraction phase instead of in the query phase. The rest is the same as H_2.

Hybrid H_4: Same as H_3, except that during the extraction phase, the challenger checks whether $\hat{C}^l(h_u^{b,itr}) \neq \mathsf{PTP.Eval}(pp, \hat{k}, h_u^{b,itr})$ holds for some b^*, l^*, itr^*, u^* where $b \in \{0,1\}$, $l \in [L]$, $u \in [d]$, $itr = l, l+1, \ldots, L$ and $\hat{C}^l$ is the l-th watermarked circuit from the challenger. If there exist such b^*, l^*, itr^*, u^*, then the experiment aborts and outputs Bad_2. The rest is the same as H_3.

Hybrid H_5: Same as H_4, except that during the extraction phase, the challenger checks whether $\tilde{C}^{\tilde{l}}(h_u^{b,itr}) \neq \mathsf{PTP.Eval}(pp, \hat{k}, h_u^{b,itr})$ holds for some b^*, l^*, itr^*, u^* where $b \in \{0,1\}$, $u \in [d]$, $itr = \tilde{l}, \tilde{l}+1, \ldots, L$, $\tilde{l}$ is the level number and $\tilde{C}^{\tilde{l}}$ is the $\tilde{l}$-th watermarked circuit from the adversary. If there exist such b^*, l^*, itr^*, u^*, then abort the experiment and output Bad_3. Otherwise, set $\tilde{X}^{b,itr} = \hat{X}^{b,itr}$ for $itr = \tilde{l}, \tilde{l}+1, \ldots, L$ and $b \in \{0,1\}$. The rest is the same as H_4.

Hybrid H_6: Same as H_5, except that during the extraction phase, for the level number $\tilde{l}$ from the adversary, re-define $ctr_i^{\tilde{l}} = \|\{j|\tilde{C}(x_{ij}^{0,\tilde{l}}) = \hat{C}^{\tilde{l}-1}(x_{ij}^{0,\tilde{l}})\}\|$ for $i \in [I]$. The rest is the same as H_5.

Hybrid H_7: Same as H_6, except that when the challenger responds to the challenge oracle, it uses different and uniformly sampled $\{\eta_u^{b,l}\}_{u \in [d]}^{b \in \{0,1\}, l \in [L]}$. The rest is the same as in H_6.

Hybrid H_8: Same as H_7, except that during the extraction phase, the challenger aborts the experiment and outputs Bad_4 if there exist $b \in \{0,1\}$, $i, i' \in [I], j, j' \in [J], l, l' \in [L]$ such that $(i,j,l) \neq (i',j',l')$ but $\hat{x}_{ij}^{b,l} = \hat{x}_{i'j'}^{b,l'}$. The rest is the same as H_7.

Lemma 1. *If Π_{PRF} is secure, then for all efficient adversaries $\mathcal{A}$,*

$$|\Pr[H_0(\mathcal{A}) \neq m^{\tilde{l}}] - \Pr[H_1(\mathcal{A}) \neq m^{\tilde{l}}]| = \mathrm{negl}(\lambda).$$

Proof. Any adversary who can distinguish H_0 and H_1 with non-negligible advantage can be used to break the security of the PRF.

Lemma 2. *If Π_{PTP} is key-injective, then for all efficient adversaries,*

$$|\Pr[H_1(\mathcal{A}) \neq m^{\tilde{l}}] - \Pr[H_2(\mathcal{A}) \neq m^{\tilde{l}}]| = \mathrm{negl}(\lambda).$$

Proof. H_1 and H_2 are identical if H_2 does not output Bad_1. In the following, we prove that Bad_1 happens with a negligible probability. If there exists an $l \in [L]$, such that $Y_{k_1}^{0,l} = Y_{k_2}^{0,l}$ or $Y_{k_1}^{1,l} = Y_{k_2}^{1,l}$ for two different keys k_1 and k_2 queried by the adversary at the l-th level, then $\mathsf{PTP.Eval}(pp, k_1, h_u^{b,l}) = \mathsf{PTP.Eval}(pp, k_2, h_u^{b,l})$ for all $u \in [d]$ and some $b \in \{0,1\}$, which happens with a negligible probability due to the key-injectivity of Π_{PTP}.

Lemma 3. *If Π_{PTP} satisfies selective constrained pseudorandomness, then for all efficient adversaries $\mathcal{A}$,*

$$|\Pr[H_2(\mathcal{A}) \neq m^{\tilde{l}}] - \Pr[H_3(\mathcal{A}) \neq m^{\tilde{l}}]| = \mathrm{negl}(\lambda).$$

Proof. $\mathbf{H}_2$ and $\mathbf{H}_3$ are identical if the adversary never makes a query on the key $\hat{k}$ sampled by the challenger answering the challenge oracle. If there exists an adversary $\mathcal{A}$ that can distinguish $\mathbf{H}_2$ and $\mathbf{H}_3$ with a non-negligible advantage ϵ, then an adversary $\mathcal{B}$ can be constructed from $\mathcal{A}$ to break the selective constrained pseudorandomness of Π_{PTP}.

1. First, $\mathcal{B}$ samples $T = \{x_{ij}^l\}_{i\in[I],j\in[j]}^{l\in[L]} \leftarrow \{0,1\}^n$ uniformly at random and it sends T to the challenger that simulates the scheme Π_{PTP}. Next, the challenger runs $(pp,tk) \leftarrow$ PTP.Setup(1^λ), $msk \leftarrow$ PTP.SampleKey(pp) and $sk_T \leftarrow$ PTP.PCst(pp,msk,T). Then, $\mathcal{B}$ receives pp and a circuit $C(\cdot) =$ PTP.PCstEval$(pp,sk_T,\cdot)$ from the challenger.

2. $\mathcal{B}$ simulates $\mathbf{H}_2$ and $\mathbf{H}_3$ for the adversary $\mathcal{A}$. It sends pp from the challenger to the adversary $\mathcal{A}$. The remaining setup is the same as in $\mathbf{H}_2$ and $\mathbf{H}_3$.

3. During the query phase, $\mathcal{B}$ answers the marking queries at the first level as in $\mathbf{H}_2$ and $\mathbf{H}_3$. For marking oracle queries at l-th level ($l \neq 1$), since $\mathcal{B}$ cannot receive the testing key tk from the challenger, it computes $ctr_i^v = \|\{j|C^{l-1}(x_{ij}^{0,v}) = $ PTP.PCstEval$(pp,sk^{l-1},x_{ij}^{0,v})\}\|$ for $v = l, l+1, \ldots, L$, $i \in [I]$ and $j \in [J]$, where $C^{l-1}(\cdot)$ is the marked circuit from the adversary $\mathcal{A}$ and sk^{l-1} is the deciphered constraint key. The rest of marking procedure remains the same. When $\mathcal{A}$ accesses challenge oracle, $\mathcal{B}$ returns $\{C^l(\cdot) =$ PTP.PCstEval$(pp,sk_T,\cdot)\}_{l\in[L]}$ to the adversary.

4. Let $k_1, k_2, \ldots, k_Q \in \mathcal{K}$ be the keys queried by $\mathcal{A}$ and Q be the maximum query number. At the end of the query phase, $\mathcal{B}$ chooses an index $i \xleftarrow{\$} [Q]$ uniformly at random and computes $y =$ PTP.Eval(pp,k_i,x_{11}^1) where $x_{11}^1 \in T$. Then, it makes a query x_{11}^1 to the challenger for the selective constrained pseudorandomness and receives a response $\hat{y}$. If $y = \hat{y}$, then $\mathcal{B}$ outputs 1; otherwise, it outputs 0.

Since $\mathcal{A}$ can distinguish $\mathbf{H}_2$ from $\mathbf{H}_3$ with a non-negligible probability ϵ, then with the same probability, it submits a PRF key which is exactly the same key sampled by the challenger. Now, consider the following two cases:

- Suppose that for the query x_{11}^1, the challenger for the constrained pseudorandomness experiment answers PTP.Eval(pp,msk,x_{11}^1). With probability ϵ/Q, $k_i = msk$ where k_i is the key queried by $\mathcal{A}$ but chosen by $\mathcal{B}$. In this case, $y = \hat{y}$ and $\mathcal{B}$ outputs 1 with probability at least ϵ/Q.
- Suppose the challenger for the constrained pseudorandomness experiment answers a truly random value. Then, $y = \hat{y}$ with a probability $\frac{1}{2^m}$ which is negligible.

To sum up, $\mathcal{B}$ can break the constrained pseudorandomness of Π_{PTP} with an advantage $\epsilon/Q - \frac{1}{2^m}$, where ϵ is non-negligible, Q is polynomial in λ. Thus, $\mathbf{H}_2$ and $\mathbf{H}_3$ are indistinguishable under the condition that Π_{PTP} is selectively constrained pseudorandom.

Lemma 4. *If Π_{PTP} satisfies selective evaluation correctness, then for all adversaries $\mathcal{A}$,*

$$|\Pr[\mathbf{H}_3(\mathcal{A}) \neq m^{\tilde{l}}] - \Pr[\mathbf{H}_4(\mathcal{A}) \neq m^{\tilde{l}}]| = negl(\lambda).$$

Proof. Hybrids $\mathbf{H}_3$ and $\mathbf{H}_4$ are identical only if in $\mathbf{H}_4$, the challenger does not output Bad_2. For all $l \in [L]$, $\hat{C}^l(\cdot) = \mathsf{PTP.PCstEval}(pp, sk_T^l, \cdot)$. Since all $\{h_u^{0,l}, h_u^{1,l}\}_{u \in [d]}^{l \in [L]}$ are sampled uniformly at random and independent of other parameters, and Π_{PTP} satisfies selective evaluation correctness, $\Pr[\hat{C}^l(h_u^{b,itr}) \neq \mathsf{PTP.Eval}(pp, \hat{k}, h_u^{b,itr})] = \mathrm{negl}(\lambda)$ for $b \in \{0,1\}$, $itr = \tilde{l}, \tilde{l}+1, \ldots, L$ and $u \in [d]$. Since $L = \omega(\log \lambda)$ and $d = \mathrm{poly}(\lambda)$, Bad_2 is output in $\mathbf{H}_4$ with negligible probability by a union bound. Thus, Hybrids $\mathbf{H}_3$ and $\mathbf{H}_4$ are indistinguishable.

Lemma 5. *For all unremoving-admissible adversary $\mathcal{A}$,*

$$|\Pr[\mathbf{H}_4(\mathcal{A}) \neq m^{\tilde{l}}] - \Pr[\mathbf{H}_5(\mathcal{A}) \neq m^{\tilde{l}}]| = \mathrm{negl}(\lambda).$$

Proof. We prove that the output distributions of $\mathbf{H}_4$ and $\mathbf{H}_5$ are statistically indistinguishable. In the following, firstly prove that Bad_3 in $\mathbf{H}_5$ is output by the challenger with negligible probability; then, prove that with high probability, $\tilde{Y}^{b,itr} = \hat{Y}^{b,itr}$ for $b \in \{0,1\}$, $itr = \tilde{l}, \tilde{l}+1, \ldots, L$.

- Note that $\{h_u^{0,l}, h_u^{1,l}\}_{u \in [d]}^{l \in [L]}$ do not relate to the challenger's behavior and the adversary's view until the extraction phase. The sampling of $\{h_u^{0,l}, h_u^{1,l}\}_{u \in [d]}^{l \in [L]}$ can be deterred at the extraction phase. Since the adversary is unremoving-admissible, $\tilde{C} \sim_f \hat{C}^{\tilde{l}}$ where $\frac{1}{f} = \mathrm{negl}(\lambda)$ and $\tilde{l}$ is the level number from the adversary at the challenge phase. Since all $\{h_u^{0,l}, h_u^{1,l}\}_{u \in [d]}^{l \in [L]}$ are sampled uniformly and independent of $\tilde{C}$ and $\hat{C}^{\tilde{l}}$, for $b \in \{0,1\}$, $\Pr[\tilde{C}(h_u^{b,l}) \neq \hat{C}^{\tilde{l}}(h_u^{b,l})] \leq \frac{1}{f} = \mathrm{negl}(\lambda)$. Besides, since $L = \omega(\log \lambda)$ and $d = \mathrm{poly}(\lambda)$, by a union bound, for all $b \in \{0,1\}, l \in [L], u \in [d]$, $\Pr[\tilde{C}(h_u^{b,l}) = \hat{C}^{\tilde{l}}(h_u^{b,l})] \geq 1 - \mathrm{negl}(\lambda)$. If Bad_2 in $\mathbf{H}_4$ is not output, then $\hat{C}^{\tilde{l}}(h_u^{b,itr}) = \mathsf{PTP.Eval}(pp, \hat{k}, h_u^{b,itr})$ for $b \in \{0,1\}, itr = \tilde{l}, \tilde{l}+1, \ldots, L, u = 1, 2, \ldots, d$. Hence, the Bad_3 in $\mathbf{H}_5$ is output by the challenger with negligible probability.
- As discussed above, Bad_3 in $\mathbf{H}_5$ is output by the challenger with negligible probability. In other words, $\tilde{y}_u^{b,itr} = \tilde{C}(h_u^{b,itr}) = \mathsf{PTP.Eval}(pp, \hat{k}, h_u^{b,itr}) = \hat{y}_u^{b,itr}$ with high probability for $b \in \{0,1\}, itr = \tilde{l}, \tilde{l}+1, \ldots, L$ and $u \in [d]$. Hence, in both $\mathbf{H}_4$ and $\mathbf{H}_5$, $\tilde{X}^{b,itr} = \hat{X}^{b,itr}$ for $b \in \{0,1\}, itr = \tilde{l}, \tilde{l}+1, \ldots, L$.

Lemma 6. *If Π_{PTP} satisfies selective verification correctness, then for all efficient and unremoving-admissible adversaries $\mathcal{A}$,*

$$|\Pr[\mathbf{H}_5(\mathcal{A}) \neq m^{\tilde{l}}] - \Pr[\mathbf{H}_6(\mathcal{A}) \neq m^{\tilde{l}}]| = \mathrm{negl}(\lambda).$$

Proof. Since Bad_1, Bad_2, Bad_3 do not happen, $\tilde{X}^{b,itr} = \hat{X}^{b,itr}$ for $b \in \{0,1\}, itr = \tilde{l}, \tilde{l}+1, \ldots, L$ where $\tilde{l}$ is the level number output by the adversary. By unremoving-admissibility of the adversary $\mathcal{A}$, with high probability $\tilde{C}(x_{ij}^{b,itr}) = \hat{C}^{\tilde{l}}(x_{ij}^{b,itr})$ for $b \in \{0,1\}, itr = \tilde{l}, \tilde{l}+1, \ldots, L, i \in [I]$ and $j \in [J]$. Then, by the verification correctness,

$$\mathsf{PTP.Test}(pp, tk, \tilde{C}(x_{ij}^{0,itr}))$$
$$= \mathsf{PTP.Test}(pp, tk, \hat{C}^{\tilde{l}}(x_{ij}^{0,itr})) = \begin{cases} 1 & i = 1, 2, \ldots, m^{\tilde{l}} \\ 0 & i = m^{\tilde{l}} + 1, m^{\tilde{l}} + 2, \ldots, L. \end{cases}$$

Thus, the counter $ctr_i^{\tilde{l}}$ are computed the same in $\mathbf{H_5}$ and $\mathbf{H_6}$. $\mathbf{H_5}$ and $\mathbf{H_6}$ are indistinguishable.

We first prove the indistinguishability between $\mathbf{H_7}$ and $\mathbf{H_8}$. Then, the indistinguishability between $\mathbf{H_6}$ and $\mathbf{H_7}$ is proven.

Lemma 7. *For all efficient and unremoving-admissible adversaries $\mathcal{A}$,*

$$|\Pr[\mathbf{H_7}(\mathcal{A}) \neq m^{\tilde{l}}] - \Pr[\mathbf{H_8}(\mathcal{A}) \neq m^{\tilde{l}}]| = \mathrm{negl}(\lambda).$$

Proof. The difference between Hybrids $\mathbf{H_7}$ and $\mathbf{H_8}$ is the event Bad_4. Since the probability of Bad_4 happening is $\frac{(IJL)^2}{2^{n-1}}$ which is negligible since $I, J, L = \omega(\log \lambda)$ and $n = \mathrm{poly}(\lambda)$, Hybrids $\mathbf{H_7}$ and $\mathbf{H_8}$ are indistinguishable.

Next, prove that Hybrids $\mathbf{H_8}$ outputs $m^{\tilde{l}}$ with non-negligible probability. First, prove that with high probability, $ctr_i^{\tilde{l}} = J$ for adversary's level number $\tilde{l}$ and $i \in [m_{\tilde{l}}]$. Since the adversary $\mathcal{A}$ is unremoving-admissible, for a negligible function $\frac{1}{f(n)}$, $\tilde{C}(\cdot) \sim_f \hat{C}^{\tilde{l}}(\cdot)$ where $\tilde{C}(\cdot)$ is the challenge response circuit from the adversary and $\hat{C}^{\tilde{l}}(\cdot)$ is the challenge circuit watermarked at the $\tilde{l}$-th level. Since $\{\hat{x}_{ij}^{b,l}\}_{i \in [I], j \in [J]}^{b \in \{0,1\}, l \in [L]}$ used for answering the challenge query are sampled uniformly and independent of the adversary's view, $\tilde{C}(\hat{x}_{ij}^{b,l}) = \hat{C}^{\tilde{l}}(\hat{x}_{ij}^{b,l})$ for $b \in \{0,1\}, l \in [L], i \in [I], j \in [J]$ with high probability.

Then, we prove that for any $i = m^{\tilde{l}} + 1, m^{\tilde{l}} + 2, \ldots, I$, $|ctr_i^{\tilde{l}} - ctr_{i+1}^{\tilde{l}}| \leq \frac{J}{I+1}$. Define $\overline{X}^{\tilde{l}} = \{x_{ij}^{0,\tilde{l}}\}$ where $i = m^{\tilde{l}} + 1, m^{\tilde{l}} + 2, \ldots, I$ and $j \in [J]$ and denote the size of $\overline{X}^{\tilde{l}}$ by g. Define $X_{and} = \{x | x \in \overline{X}^{\tilde{l}} \wedge \tilde{C}(x) = \hat{C}^{\tilde{l}-1}(x)\}$ and denote the size of X_{and} by u. Since the exact partition of $\overline{X}^{\tilde{l}}$ is independent of the view of the adversary $\mathcal{A}$, the distribution of $ctr_i^{\tilde{l}}$ for $i = m^{\tilde{l}} + 1, m^{\tilde{l}} + 2, \ldots, I$ is the hypergeometric distribution $\mathcal{H}(u, g, J)$. Therefore,

$$\Pr[ctr_i^{\tilde{l}} \geq (\frac{u}{g} + \frac{1}{2(I+1)})J] \leq e^{-\frac{J}{2(I+1)^2}}, \ \Pr[ctr_{i+1}^{\tilde{l}} \leq (\frac{u}{g} - \frac{1}{2(I+1)})J] \leq e^{-\frac{J}{2(I+1)^2}},$$

which are both negligible. By the union bound, the probability that there exists $i = m^{\tilde{l}} + 1, m^{\tilde{l}} + 2, \ldots, I$ such that $|ctr_i^{\tilde{l}} - ctr_{i+1}^{\tilde{l}}| \geq \frac{J}{I+1}$ is negligible. Thus, the smallest subscribe such that $|ctr_i^{\tilde{l}} - ctr_{i+1}^{\tilde{l}}| \geq \frac{J}{I+1}$ is $m_{\tilde{l}}$ with high probability.

Lemma 8. *If Π_{PTP} satisfies selectively consistent privacy, then for all efficient adversaries $\mathcal{A}$,*

$$|\Pr[\mathbf{H_6}(\mathcal{A}) \neq m^{\tilde{l}}] - \Pr[\mathbf{H_7}(\mathcal{A}) \neq m^{\tilde{l}}]| = \mathrm{negl}(\lambda).$$

Proof. Suppose that an adversary $\mathcal{A}$ can distinguish $\mathbf{H_6}$ and $\mathbf{H_7}$ with a non-negligible probability, then an adversary $\mathcal{B}$ can be constructed to break the selectively consistent privacy of the Π_{PTP}. The reduction proceeds as follows:

1. To start, $\mathcal{B}$ guesses what L messages the adversary $\mathcal{A}$ is intended to embed in the challenge phase. Suppose these L messages are guessed to be $\{m_1, m_2, \ldots, m_L\}$. Next, $\mathcal{B}$ samples two point sets T_0, T_1 with a special form uniformly at random. More specifically, if we define $T_b^{m_1} = \{x_{ij}^{bl} : x_{ij}^{bl} \xleftarrow{\$} \{0,1\}^n, \forall l \in [L], i \in [I], j \in [J]\}$, $\bar{X}_b^{m_l} = \{x_{ij}^{bl} \xleftarrow{\$} \{0,1\}^n : i \in \{m_l + 1, m_l + 2, \ldots, I - 1, I\}, j \in [J]\}$, and $X_b^{m_l} = \{x_{ij}^{bl} \in T_b^{m_{l-1}} : i \in \{m_l + 1, m_l + 2, \ldots, I - 1, I\}, , j \in [J]\}$, then for $l = 2, 3, \ldots, L$ and $b = \{0, 1\}$, $T_b^{m_l} = (T_b^{m_{l-1}} \setminus X_b^{m_l}) \cup \bar{X}_b^{m_l}$ [13]. Then, $T_0 = \{T_0^{m_1}, T_0^{m_2}, \ldots, T_0^{m_L}\}$ and $T_1 = \{T_1^{m_1}, T_1^{m_2}, \ldots, T_1^{m_L}\}$ are sent to the challenger $\mathcal{C}$.

2. $\mathcal{C}$ samples a bit β uniformly at random. Then, $\mathcal{C}$ runs the setup algorithm of the scheme Π_{PTP} and generates L constraint keys $\{sk_l\}_{l \in [L]}$ punctured at T_β. Finally, the public parameters pp and L circuits $\{C_l = \mathsf{PTP.PCstEval}(pp, sk_l, \cdot)\}_{l \in [L]}$ are sent to $\mathcal{B}$.

3. $\mathcal{B}$ invokes $\mathcal{A}$. To simulate the unremovability experiment, $\mathcal{B}$ proceeds the setup as in the watermarking scheme. At the end of the setup phase, $\mathcal{B}$ sends pp to $\mathcal{A}$.

4. In the query phase, $\mathcal{B}$ answers the queries as follows:
   - Marking oracle: There exists one difference when $\mathcal{B}$ answers the marking oracle. Since $\mathcal{B}$ does not have the testing key, $\mathcal{B}$ cannot compute the counters same as in the third step of $\mathsf{WM.Mark}_l(\cdot)$. To overcome this difficulty, $\mathcal{B}$ computes the counters by $ctr_i^{b,itr} = \sum_{j=1}^{J} 1_{\neq}(C^{l-1}(x_{ij}^{b,itr}) \neq \mathsf{PTP.Eval}(pp, k, x_{ij}^{b,itr}))$ for $itr = l, l+1, \ldots, L$ where $1_{\neq}$ is an indicator function, i.e.,

$$1_{\neq}(expression) = \begin{cases} 1 & expression \text{ is true} \\ 0 & expression \text{ is false.} \end{cases}$$

   - Challenge oracle: On input a set of challenge messages $\{m_l\}_{l=1}^{L}$, $\mathcal{B}$ checks whether it has made a correct guess. If yes, then $\mathcal{B}$ sends L circuits $\{C_l\}_{l \in [L]}$ to $\mathcal{A}$ directly. If no, then $\mathcal{B}$ aborts the experiment and outputs a bit uniformly at random.

5. $\mathcal{A}$ outputs a circuit $\tilde{C}^{\tilde{l}}$ and a level number $\tilde{l}$ when it makes no more queries. Then, $\mathcal{B}$ extracts the watermarked message from $\tilde{C}^{\tilde{l}}$. If the extracted message is not $m^{\tilde{l}}$, then $\mathcal{B}$ outputs 1; otherwise, it outputs 0.

As in Lemma 7, $\mathbf{H}_7$ and $\mathbf{H}_8$ are indistinguishable and $\mathbf{H}_8$ does not output $m^{\tilde{l}}$ with a negligible probability. Thus, it is concluded that $\mathbf{H}_7$ does not output $m^{\tilde{l}}$ with a negligible probability. By contradiction, assume that $\mathbf{H}_6$ does not output $m^{\tilde{l}}$ with a noticeable probability ϵ. In the following, we discuss two cases: $\beta = 0$ and $\beta = 1$.

   - $\beta = 0$: $\mathcal{B}$ simulates $\mathbf{H}_6$ for $\mathcal{A}$. Under our assumption, $\mathcal{B}$ outputs 1 with a probability at $\frac{1}{2} + \frac{1}{I^L}\epsilon$.

---

[13] $T_b^{m_l}$ is the puncture point set encoding the watermarking messages $m_1, m_2, \ldots, m_l$..

– $\beta = 1$: $\mathcal{B}$ simulates $\mathbf{H}_7$ for $\mathcal{A}$. Based on our proof, $\mathcal{B}$ outputs 1 with a probability at $\frac{1}{2}$ plus a negligible probability.

In our scheme, L is set to be a constant and I is a polynomial in λ. To conclude, $\mathcal{B}$ breaks the selectively consistent privacy of Π_{PTP} with a non-negligible probability which is a contradiction.

Combining all these lemmas, unremovability is proven.

B.3   Proof of Unforgeability

Proof. To start with, define the following hybrids:

Hybrid $\mathbf{H}_i$ ($i = 0, 1, 2, 3$): It is almost identical to $\mathbf{H}_i$ defined in proving unremovability, except that there is no challenge oracle. Besides, in the extraction phase of **Hybrid $\mathbf{H}_3$**, the challenger computes $\tilde{Y}^{b,l} = (\tilde{C}(h_1^{b,l}), \ldots, \tilde{C}(h_d^{b,l}))$ for $b \in \{0,1\}$ and aborts the experiment if for some k queried by the adversary at the query phase, $\tilde{Y}^{b,l} = (\mathsf{PTP.Eval}(pp, k, h_1^{b,l}), \ldots, \mathsf{PTP.Eval}(pp, k, h_d^{b,l}))$ for $b = 0, 1$. Otherwise, it proceeds as $\mathbf{H}_2$.

Lemma 9. *If Π_{PRF} is a secure PRF and Π_{PTP} is key-injective, then for all adversaries,*

$$|\Pr[\mathbf{H}_i(\mathcal{A}) \neq \perp] - \Pr[\mathbf{H}_{i+1}(\mathcal{A}) \neq \perp]| = \mathsf{negl}(\lambda), \text{for } i = 0, 1.$$

Proof. The proof follows the same arguments for Lemmas 1 and 2.

Lemma 10. *If Π_{PTP} satisfies evaluation correctness, then for all δ-unforging-admissible adversaries $\mathcal{A}$ where $\delta = \frac{1}{\mathsf{poly}(\lambda)}$,*

$$|\Pr[\mathbf{H}_2(\mathcal{A}) \neq \perp] - \Pr[\mathbf{H}_3(\mathcal{A}) \neq \perp]| = \mathsf{negl}(\lambda).$$

Proof. If $\mathbf{H}_3$ does not abort the experiment, then Hybrid $\mathbf{H}_2$ and $\mathbf{H}_3$ are statistically indistinguishable. In the following, we prove the abortion in $\mathbf{H}_3$ happens with a negligible probability.

For $l \in [L]$ and $q_l \in [Q_l]$, let $S_{q_l}^l$ be the set of points at which the circuit $\tilde{C}$ output by the adversary and the circuit computing $\mathsf{PTP.Eval}(pp, k_{q_l}^l, \cdot)$ disagree. Note that $\mathsf{PTP.Eval}(pp, k_{q_l}^l, \cdot)$ agrees at all but a negligible fraction of the whole domain with $C_{q_l}^l(\cdot)$. Here, $C_{q_l}^l$ is the marked circuit for the PRF key $k_{q_l}^l$ at the l-th level for the q_l-th query. Due to the δ-unforging-admissibility, $\frac{|S_{q_l}^l|}{2^n} \geq \delta$. Since the marking phase does not depend on $\{h_u^{0,l}, h_u^{1,l}\}_{u \in [d]}^{l \in [L]}$, the sampling of $\{h_u^{0,l}, h_u^{1,l}\}_{u \in [d]}^{l \in [L]}$ can be deterred until the extraction phase. Since each $h_u^{b,l}$ is sampled uniformly and independently, for $b \in \{0,1\}$, $l \in [L]$, $u \in [d]$ and $q_l \in [Q_l]$, we have that $\Pr[h_u^{b,l} \in S_{q_l}^l] = \frac{|S_{q_l}^l|}{2^n} \geq \delta$. Then, for all $l \in [L]$ and $q_l \in [Q_l]$, $b \in \{0,1\}$,

$$\Pr[\forall u \in [d] : h_u^{b,l} \notin S_{q_l}^l] = (1 - \frac{|S_{q_l}^l|}{2^n})^d \leq (1 - \delta)^{\lambda/\delta} \leq e^{-\lambda}, \tag{1}$$

where $d = \lambda/\delta$ and $\delta = 1/\text{poly}(\lambda)$. Since we set $\sum_{l=1}^{L} q_l = \text{poly}(\lambda)$, with negligible probability, $\mathbf{H}_3$ aborts the experiment. Thus, Hybrid $\mathbf{H}_2$ and $\mathbf{H}_3$ are statistically indistinguishable.

Lemma 11. *For all adversaries,* $\Pr[\mathbf{H}_3(\mathcal{A}) \neq \bot] = \text{negl}(\lambda)$.

Proof. Since $\mathbf{H}_3$ does not abort, then $X = \{x_{ij}^l \leftarrow \{0,1\}^n : \text{for all } l \in [L], i \in [I], j \in [J]\}$.

Since $\frac{LIJ}{2^n}$ is negligible, $\Pr[\mathsf{PTP.Test}(pp, tk^{\bar{l}}, \tilde{C}(x_{ij}^{\bar{l}})) = 1] = \frac{LIJ}{2^n} = \text{negl}(\lambda)$. By a union bound, $\Pr[ctr_i^{\bar{l}} = \sum_{j\in[J]} \mathsf{PTP.Test}(pp, tk^{\bar{l}}, \tilde{C}(x_{ij}^{\bar{l}})) = 0] = (1 - \frac{LIJ}{2^n})^J \sim$ $1 - \text{negl}(\lambda)$ for all $i \in [I]$. Thus, with high probability, 0 is extracted from $\tilde{C}$ which leads to output $\bot$ for the experiment.

Combing all these lemmas, the watermarking scheme satisfies unforgeability.

References

1. Ajtai, M.: Generating hard instances of the short basis problem. In: Wiedermann, J., van Emde Boas, P., Nielsen, M. (eds.) ICALP 1999. LNCS, vol. 1644, pp. 1–9. Springer, Heidelberg (1999). https://doi.org/10.1007/3-540-48523-6_1
2. Alwen, J., Peikert, C.: Generating shorter bases for hard random lattices (2009)
3. Alwen, J., Peikert, C.: Generating shorter bases for hard random lattices. Theory Comput. Syst. **48**(3), 535–553 (2011)
4. Barak, B., et al.: On the (im)possibility of obfuscating programs. In: Kilian, J. (ed.) CRYPTO 2001. LNCS, vol. 2139, pp. 1–18. Springer, Heidelberg (2001). https://doi.org/10.1007/3-540-44647-8_1
5. Barak, B., et al.: On the (im) possibility of obfuscating programs. J. ACM (JACM) **59**(2), 1–48 (2012)
6. Boneh, D., et al.: Fully key-homomorphic encryption, arithmetic circuit ABE and compact garbled circuits. In: Nguyen, P.Q., Oswald, E. (eds.) EUROCRYPT 2014. LNCS, vol. 8441, pp. 533–556. Springer, Heidelberg (2014). https://doi.org/10.1007/978-3-642-55220-5_30
7. Boneh, D., Kim, S., Montgomery, H.: Private puncturable PRFs from standard lattice assumptions. In: Coron, J.-S., Nielsen, J.B. (eds.) EUROCRYPT 2017. LNCS, vol. 10210, pp. 415–445. Springer, Cham (2017). https://doi.org/10.1007/978-3-319-56620-7_15
8. Boneh, D., Lewi, K., Wu, D.J.: Constraining pseudorandom functions privately. In: Fehr, S. (ed.) PKC 2017. LNCS, vol. 10175, pp. 494–524. Springer, Heidelberg (2017). https://doi.org/10.1007/978-3-662-54388-7_17
9. Boneh, D., Waters, B.: Constrained pseudorandom functions and their applications. In: Sako, K., Sarkar, P. (eds.) ASIACRYPT 2013. LNCS, vol. 8270, pp. 280–300. Springer, Heidelberg (2013). https://doi.org/10.1007/978-3-642-42045-0_15
10. Brakerski, Z., Vaikuntanathan, V.: Lattice-based FHE as secure as PKE. In: Proceedings of the 5th Conference on Innovations in Theoretical Computer Science, pp. 1–12. ACM (2014)

11. Brakerski, Z., Vaikuntanathan, V.: Constrained key-homomorphic PRFs from standard lattice assumptions. In: Dodis, Y., Nielsen, J.B. (eds.) TCC 2015. LNCS, vol. 9015, pp. 1–30. Springer, Heidelberg (2015). https://doi.org/10.1007/978-3-662-46497-7_1
12. Chvátal, V.: The tail of the hypergeometric distribution. Discret. Math. **25**(3), 285–287 (1979)
13. Cohen, A., Holmgren, J., Nishimaki, R., Vaikuntanathan, V., Wichs, D.: Watermarking cryptographic capabilities. SIAM J. Comput. **47**(6), 2157–2202 (2018)
14. Gentry, C., Peikert, C., Vaikuntanathan, V.: Trapdoors for hard lattices and new cryptographic constructions. In: Proceedings of the Fortieth Annual ACM Symposium on Theory of Computing, pp. 197–206. ACM (2008)
15. Gentry, C., Sahai, A., Waters, B.: Homomorphic encryption from learning with errors: conceptually-simpler, asymptotically-faster, attribute-based. In: Canetti, R., Garay, J.A. (eds.) CRYPTO 2013. LNCS, vol. 8042, pp. 75–92. Springer, Heidelberg (2013). https://doi.org/10.1007/978-3-642-40041-4_5
16. Gorbunov, S., Vaikuntanathan, V., Wee, H.: Predicate encryption for circuits from LWE. In: Gennaro, R., Robshaw, M. (eds.) CRYPTO 2015. LNCS, vol. 9216, pp. 503–523. Springer, Heidelberg (2015). https://doi.org/10.1007/978-3-662-48000-7_25
17. Goyal, R., Kim, S., Manohar, N., Waters, B., Wu, D.J.: Watermarking public-key cryptographic primitives. In: Boldyreva, A., Micciancio, D. (eds.) CRYPTO 2019. LNCS, vol. 11694, pp. 367–398. Springer, Cham (2019). https://doi.org/10.1007/978-3-030-26954-8_12
18. Kim, S., Wu, D.J.: Watermarking cryptographic functionalities from standard lattice assumptions. In: Katz, J., Shacham, H. (eds.) CRYPTO 2017. LNCS, vol. 10401, pp. 503–536. Springer, Cham (2017). https://doi.org/10.1007/978-3-319-63688-7_17
19. Kim, S., Wu, D.J.: Watermarking PRFs from lattices: stronger security via extractable PRFs. In: Boldyreva, A., Micciancio, D. (eds.) CRYPTO 2019. LNCS, vol. 11694, pp. 335–366. Springer, Cham (2019). https://doi.org/10.1007/978-3-030-26954-8_11
20. Lyubashevsky, V., Wichs, D.: Simple lattice trapdoor sampling from a broad class of distributions. In: Katz, J. (ed.) PKC 2015. LNCS, vol. 9020, pp. 716–730. Springer, Heidelberg (2015). https://doi.org/10.1007/978-3-662-46447-2_32
21. Micciancio, D., Peikert, C.: Trapdoors for lattices: simpler, tighter, faster, smaller. In: Pointcheval, D., Johansson, T. (eds.) EUROCRYPT 2012. LNCS, vol. 7237, pp. 700–718. Springer, Heidelberg (2012). https://doi.org/10.1007/978-3-642-29011-4_41
22. Naccache, D., Shamir, A., Stern, J.P.: How to copyright a function? In: Imai, H., Zheng, Y. (eds.) PKC 1999. LNCS, vol. 1560, pp. 188–196. Springer, Heidelberg (1999). https://doi.org/10.1007/3-540-49162-7_14
23. Nishimaki, R.: How to watermark cryptographic functions. In: Johansson, T., Nguyen, P.Q. (eds.) EUROCRYPT 2013. LNCS, vol. 7881, pp. 111–125. Springer, Heidelberg (2013). https://doi.org/10.1007/978-3-642-38348-9_7
24. Quach, W., Wichs, D., Zirdelis, G.: Watermarking PRFs under standard assumptions: public marking and security with extraction queries. In: Beimel, A., Dziembowski, S. (eds.) TCC 2018. LNCS, vol. 11240, pp. 669–698. Springer, Cham (2018). https://doi.org/10.1007/978-3-030-03810-6_24
25. Yang, R., Au, M.H., Lai, J., Xu, Q., Yu, Z.: Collusion resistant watermarking schemes for cryptographic functionalities. In: Galbraith, S.D., Moriai, S. (eds.) ASI-

ACRYPT 2019. LNCS, vol. 11921, pp. 371–398. Springer, Cham (2019). https://doi.org/10.1007/978-3-030-34578-5_14
26. Yang, R., Au, M.H., Yu, Z., Xu, Q.: Collusion resistant watermarkable PRFs from standard assumptions. In: Micciancio, D., Ristenpart, T. (eds.) CRYPTO 2020. LNCS, vol. 12170, pp. 590–620. Springer, Cham (2020). https://doi.org/10.1007/978-3-030-56784-2_20
27. Yoshida, M., Fujiwara, T.: Toward digital watermarking for cryptographic data. IEICE Trans. Fundam. Electron. Commun. Comput. Sci. **94**(1), 270–272 (2011)

Evaluating the Future Device Security Risk Indicator for Hundreds of IoT Devices

Pascal Oser[1,2]([✉]) [iD], Felix Engelmann[3] [iD], Stefan Lüders[1] [iD], and Frank Kargl[2] [iD]

[1] European Organization for Nuclear Research (CERN), Geneva, Switzerland
stefan.lueders@cern.ch
[2] Ulm University, Ulm, Germany
{pascal.oser, frank.kargl}@uni-ulm.de
[3] IT University of Copenhagen, Copenhagen, Denmark
fe-research@nlogn.org

Abstract. IoT devices are present in many, especially corporate and sensitive, networks and regularly introduce security risks due to slow vendor responses to vulnerabilities and high difficulty of patching. In this paper, we want to evaluate to what extent the development of future risk of IoT devices due to new and unpatched vulnerabilities can be predicted based on historic information. For this analysis, we build on existing prediction algorithms available in the SAFER framework (prophet and ARIMA) which we evaluate by means of a large data-set of vulnerabilities and patches from 793 IoT devices. Our analysis shows that the SAFER framework can predict a correct future risk for 91% of the devices, demonstrating its applicability. We conclude that this approach is a reliable means for network operators to efficiently detect and act on risks emanating from IoT devices in their networks.

Keywords: IoT · Security risk assessment · Device identification · Firmware analysis · Vulnerability analysis · Risk prediction · Future risk · SAFER network

1 Introduction

IoT devices are becoming more wide-spread in areas such as smart homes, smart cities, but also in research and office environments. Their sheer number, heterogeneity and limited patch availability provide significant challenges for the security of local networks and the internet in general. This stems from the observation that many devices have vulnerabilities and the availability of patches varies greatly by device and vendor. The systematic evaluation of device risks, which is essential for mitigation decisions, is currently a skill-intensive task that requires expertise like network vulnerability scanning, or even manual firmware binary analysis.

This paper presents an in-depth and large-scale IoT device security assessment by utilizing the risk *prediction & scoring* component of the Security Assessment Framework for Embedded-device Risks (SAFER) [23]. SAFER is a highly-automated framework to identify devices on the network for estimating their

G. Lenzini and W. Meng (Eds.): STM 2022, LNCS 13867, pp. 52–70, 2023.
https://doi.org/10.1007/978-3-031-29504-1_3

current as well as future security risk based on publicly retrievable firmware information. For this work, we focus on the future prediction quality as the current risk of a device is calculated deterministically. To estimate device risks, we rely on information from among others public vulnerability databases, vendor published software license statements and firmware release notes. Moreover, based on past vulnerability data and vendor patch intervals for device models, SAFER's risk component extrapolates those observations into the future using different predefined and automatically parameterized prediction models. This lets SAFER estimate an indicator for future device security risks enabling users to be aware of devices exposing high risks in the future.

One major strength of using SAFER's risk component over other approaches is the ability to perform significantly automated risk assessments for risks associated to the current firmware, the detection of already patched vulnerabilities and the estimation of a future device risk indicator based on past observed and estimated future information.

Oser et al. [23] describe SAFER extensively, however the evaluation focuses on device identification and the authors only performed a preliminary evaluation of the risk metrics with 38 device models. They concluded that for their reduced data-set, current and future risk was predicted with almost 100% accuracy. As this perfect prediction could be caused by a too small data sample, it remains to show the prediction performance in a large scale, realistic setting.

To investigate this strength, we deployed a version of SAFER with an enhanced risk *prediction & scoring* component in the network of a large multinational organization to systematically assess the security level for hundreds of IoT devices in large-scale networks. In this work, we utilized SAFER to estimate the risks of 6,123 different firmware versions for 838 device models. The future security risk is calculated from a patch trend, indicating how long vendor needs to patch vulnerabilities, and a vulnerability trend, indicating the likely severity of future vulnerabilities. For the combined future device security risks, SAFER achieved correct predictions for 91.30% of 793 device models using vendor published information. This shows that the preliminary evaluation lacked in depth and that SAFER is indeed a valuable tool in realistic settings with many devices and can guide administrators to identify devices that are likely to cause security problems in the future.

2 Related Work

To give an overview of the field, we first introduce related work for the separate aspects of our contribution like vulnerability prediction and risk scoring, as we are not aware of a similar combined work.

Vulnerability Prediction. Wu et al. [36] use a multi-variable Long Short-Term Memory (LSTM) to predict time-series data for vulnerabilities. The authors use browser vulnerabilities from May 2008 to May 2019 of the National Vulnerability Database (NVD) to retrieve vulnerabilities for five web browsers which they use to evaluate their approach on. An Auto-regressive Integrated Moving Average (ARIMA) model is also trained on this data to compare the results with the proposed solution. The authors state that their solution predicts the number of vulnerabilities for Chrome better than an ARIMA model and achieve an RMSE of 8.032 by their approach and an RMSE of 15.210 by using an ARIMA model. The RMSE is based on the error of the predicted vulnerabilities for 40% of the browser data-set.

ReVeal by Chakraborty et al. [4] performs vulnerability prediction on the Linux Debian Kernel and Chromium. The authors use these code bases because those are well-maintained public projects with large evolutionary history including plenty of publicly available vulnerability reports. They compare ReVeal with four other approaches [18, 19, 28, 39] of related work and state that existing works have the following limitations: 1) they introduce data duplication, 2) do not handle data imbalance, 3) do not learn semantic information, and 4) lack class separability. The authors compare their Deep-Learning approach with the best performing model in the literature and state that ReVeal performs up to 33.57% better in precision and 128.38% better in recall.

Dam et al. [5] propose an LSTM model to learn both semantic and syntactic features of code. With this knowledge, the model predicts vulnerabilities for 18 Android applications written in Java. The authors claim a prediction improvement, compared to traditional software metrics approaches, of 3–58% for within-project prediction and 85% for cross-project prediction.

Jimenez et al. [12] analyzed the Linux Kernel with more than 570,000 commits from 2005 to 2016. They observed that approaches based on header files and function calls perform best for future vulnerability prediction. The authors state that text mining is the best technique when aiming at random instances and identified that code metrics, on the opposite, perform poorly.

Risk Assessment Approaches. In the following, we introduce risk assessment approaches partly focused on IoT. Bahizad [2] discusses the increase of IoT devices and the resulting risks a network has to deal with. Bahizad states: "[...] due to the connections between IoT devices, the security of one device is also dependent on the security of other devices and the cascading effects of its vulnerabilities to the whole system. As these devices increase, the risk added to the system increases." [2]. This observation emphasizes the problems existing in large heterogeneous networks of IoT devices and motivates the need for IoT risk assessment solutions and their performance evaluation in real world scenarios.

Shivraj et al. [30] propose a model-driven risk assessment framework based on graph theory. The authors modeled the system using attack trees and simulated different modes of attack propagation on it. Ultimately, they stress the usefulness of their work by empirical analysis and experiments on the STRIDE[1] and LINDDUN[2] threat models. The authors performed an in-depth analysis of 207 vulnerabilities to identify the time until a vulnerability was discovered and

---

[1] https://msdn.microsoft.com/en-us/library/ee823878(v=cs.20).aspx.
[2] https://www.linddun.org.

advisories released. On average, it takes 5.3 years until the investigated vulnerabilities are publicly registered. As a conclusion, they propose guidelines to improve reporting and consistency of ICS CVE information.

Li et al. [17] developed a method to detect security risks of devices based on firmware fingerprinting. The authors retrieved 9,716 firmware images from third-party websites and 347,685 security reports. For firmware fingerprinting, they identified subtle differences in firmware file-systems as well as analyzing contained HTML files after labeling them manually. Using word embeddings in combination with two-layer neural networks and regular expressions, the authors claim 91% precision and 90% recall for fingerprinting firmware images. 6,898 security reports contain firmware and vulnerability information. The authors also identified that more than 10% of detected firmware vulnerabilities do not have any patches listed in public databases.

Duan et al. [7] propose an automated security assessment framework for IoT networks. The proposed security model automatically assesses the security of the IoT network by capturing potential attack paths and identifying the most vulnerable ones. They use machine learning and natural language processing to analyze vulnerability descriptions and predict vulnerability metrics of new vulnerabilities with more than 90% accuracy. The predicted vulnerability metrics are used in a graphical security model consisting of attack graphs.

Rodríguez et al. [27] quantified how IoT manufacturers may act as "superspreaders" for device infections. The authors scanned the internet during two months for Mirai-infected devices resulting in 31,950 infected IoT devices of 70 unique manufacturers found in 68 countries. 53% of the 70 identified manufacturers offer firmware or software downloads on their websites, 43% provide password changing procedures (as Mirai targets devices using standard passwords), and 26% of the manufacturers offer advice to protect devices from attacks. In total, they identified that nine vendors share almost 50% of the infections identified.

Security Scoring Mechanisms. We introduce approaches of multiple researchers who work on specialized ways of the Common Vulnerability Scoring System (CVSS). Some propose improved versions of the CVSS [34], device type specific scoring systems [26,32], propose vulnerability assessment methods [33] or risk management for embedded devices [10]. Johnson et al. [13] compare the credibility of the CVSS scoring data of different vulnerability sources of NVD, X-Force, OSVDB, CERT-VN, and Cisco. Le and Hoang [16] propose an approach to compute the probability distribution of cloud security threats based on Markov chains and the CVSS.

3   Preliminaries

Before presenting our evaluation, we first introduce details of the SAFER framework on which our evaluation builds. We also explain what enhancements we have introduced compared to the version described in [23].

The SAFER framework implements various components to 1) identify IoT devices, 2) gather vulnerability information for identified devices, 3) score this information and 4) display results to support SAFER's users in making informed decisions regarding their devices' security.

Initiating a device assessment starts by specifying the host-name, IP address or network range which multiple identification mechanisms will scan. SAFER combines the results of different identification mechanisms using a probabilistic logic framework called Subjective Logic [14]. The next step is then to retrieve available firmware image and software license statements per firmware version from a vendor's support web-site. This lets SAFER determine contained software and additional information about the firmware. Next, the framework gathers publicly known vulnerabilities for identified device models and all contained software using public vulnerability repositories. In its last step, all information is aggregated and SAFER computes a Current Device Security Risk Indicator (CDSRI) and a Future Device Security Risk Indicator (FDSRI) based on past evidence. Previous works about SAFER describe its components [1,22,24] and the overall framework [23] but evaluate in particular the FDSRI with only few devices and less details.

Predicting a Device's Future Risk. SAFER extrapolates past observations into a prediction of a future security risk a device may pose due to future vulnerabilities. It also considers how quickly such future vulnerabilities might be patched by vendors. To calculate this so called Future Device Security Risk Indicator (FDSRI), SAFER tries to separately predict the frequency and severity level of such future vulnerabilities and also estimates future patch intervals both based on observed historic data. The FDSRI is composed of two dimensions: To predict future vulnerability severity levels we calculate a so-called Vulnerability Trend (VT). For this, SAFER takes all previously identified vulnerabilities of a device into account. The Patch Trend (PT), on the other hand, predicts future patch intervals based on time intervals of previously patched vulnerabilities; from the date they became publicly known until the vendor released a patch. For those predictions, SAFER applies various prediction models on its vulnerability severity and patch data, specifically Facebook Prophet [31], simple moving average, an auto regressive model[3] and different Auto Regressive Integrated Moving Average (ARIMA) models[4] based on the Box-Jenkins method [3].

4   Predicting the Patch Trend

This section introduces the evaluation of the *Patch Trend*. First, we introduce how we built our data-set containing dates of past patch intervals for our device

---

[3] https://www.statsmodels.org/dev/generated/statsmodels.tsa.ar_model.AutoReg.html.

[4] https://www.statsmodels.org/stable/generated/statsmodels.tsa.arima.model.ARIMA.html.

models under investigation. Second, based on the device model's past patch behavior, we evaluate different prediction models to measure how well these predict future patch intervals. Third, we discuss the predicted future patch intervals and how they compare with actually observed patch intervals to identify prediction accuracy.

4.1   Data-Set Observation

SAFER analyzed, exemplary for Axis, hundreds of release notes and license statements. Those support SAFER with detailed information about firmware-contents by simply parsing text files. The parser is vendor specific and needs to understand the semantics of patch notes. We noticed that the structure is consistent within vendors but differ slightly between them, requiring adaptation of the parser. SAFER creates a new data-set per device model containing the release dates of relevant CVEs and the time in days the vendor required to patch those[5]. The data also varies on the amount of patches the vendor applied to device models. SAFER identified that 423 device models have not been patched by their vendor compared to 370 device models having received between 2 and 19 patches. The data shows an average amount of 38.84 patches for all 793 device models and an average of 83.26 patches for the 370 device models having received at least one patch. The average interval in days the exemplary analyzed vendor Axis requires to patch registered vulnerabilities in their analyzed device models is about 953 days. The data divides in the first quartile ranging to 316 days, the second quartile ranging to 634 days and the third quartile ranging to 1,170 days. The maximum time Axis required to patch a device model's vulnerability was 6,017 days resulting in circa 16.5 years. A histogram of the years (x-axis) Axis needed for patching vulnerabilities (y-axis) over all their device models is shown in Fig. 1.

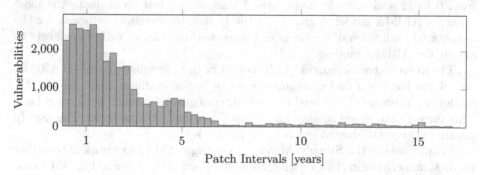

Fig. 1. Patch interval overview in years for patched vulnerabilities.

After having established our data set, we proceed by using the observed patch intervals per device model and estimate future patch intervals.

---

[5] The raw data-set will become available at https://safer.network/.

4.2   Prediction Model Evaluation

To evaluate the predicted patch intervals and, hence, derive a patch trend afterwards, we analyzed the data-sets of past patch time-spans for each device model separately.

Each of the data-sets was split in a 66% training- and 34% test-set. We then trained the device model specific prediction models on the training-set of past patch intervals and predicted future patch intervals which we compared with observed values of the test-set.

While this was also done in the original version of SAFER, we introduce a number of changes in this paper: originally, the AR model was not evaluated individually and the parameter *changepoint-prior-scale*[6] of Facebook's Prophet was set to a static value. In this work, we enhanced SAFER by performing a semi-automated data analysis and automated 1) the AR model's lags and 2) Prophet's *changepoint-prior-scale* parameter to tune parameters per device model data-set. This improvement enables the prediction model to handle different behavior regimes, e.g. if a vendor suddenly stopped patching a once frequently patched device. Per device model, we evaluated the parameter value for *changepoint_prior_scale* which adjusts the trend flexibility from rather static (0.01) to highly fluctuating (2.1) trends. We find that adding those two parameterizing approaches increased the prediction accuracy significantly.

We use **Facebook's Prophet** as one of our prediction models. There, we changed *seasonality_mode*[7] to multiplicative which performed better for PT data as seasonality rather grows with data in comparison to being a constant factor where additive models are used. In time series, seasonality refers to fluctuations that occur at regular intervals and prevent regular patterns.

For the **ARIMA** prediction model, we use *pmdarima*[8] to perform automated evaluation of the best fitting ARIMA model including others like SARIMAX. The tool evaluates all p, d, and q parameters which we configured to be in range from 0 to 12 and uses the augmented Dickey-Fuller test to evaluate the most accurate ARIMA model. We consider the parameter range as sufficient for the automated evaluation of device model data, because higher values are likely to overfit the ARIMA model.

The **auto-regressive** model (AR) is part of the previously discussed ARIMA model but has the d and q parameters set to 0. For evaluation, we use the AR model of *statsmodels*[9] and used the default configuration which performed best. This means that no seasonality and a constant trend is considered. The lags (p parameter) were evaluated individually per data-set.

Finally, we use the **Simple Moving Average** (SMA) to compare the other prediction mechanisms to a relatively simple approach for forecasting. SMA con-

---

[6] https://facebook.github.io/prophet/docs/trend_changepoints.html#automatic-changepoint-detection-in-prophet.

[7] https://facebook.github.io/prophet/docs/multiplicative_seasonality.html.

[8] https://pypi.org/project/pmdarima/.

[9] https://www.statsmodels.org/dev/generated/statsmodels.tsa.ar_model.AutoReg.html.

siders a defined amount of past data-points when forecasting which is known as
the so-called window. Calculating the average of past patch intervals in the window and dividing it by the number of patch-intervals leads to the estimated
forecast for the next patch-interval. This process is performed recursively until
reaching the end of the test-set. The window size was evaluated using the same
approach like for ARIMA.

The prediction model results are shown in Table 1. Our enhanced version of
SAFER was able to predict patch intervals for 793 device models requiring a
minimum amount of two past patch intervals to estimate future patch intervals.
If a device model did receive less than two patches, the prediction models cannot
predict future data as no trend is available and considers the PT of the device
model as *slow* as in [23].

To compare how accurate the prediction models estimated the observed patch
intervals of the test-set, we calculated the Root Mean Square Error (RMSE) and
Median Absolute Deviation (MAD). The RMSE is the square root of the average
of the square of all errors. The MAD is defined as "a measure of scale based on
the median of the absolute deviations from the median of the distribution." [11].
MAD and RMSE are important indicators when considering both normally and
not normally distributed data.

Table 1. Comparison of average patch interval
prediction error

Version	Predictor	Devices	RMSE	MAD
Our	Prophet	793	443.75	119.0
Our	ARIMA	793	417.14	154.15
Our	AR	793	53.17	25.13
Our	SMA	793	166.52	65.08
Original	Prophet	38	-	14.31
Original	ARIMA	38	-	24.12

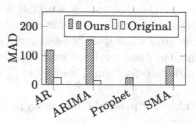

Calculating the median over all AR device model prediction errors results in
an RMSE of 53.17 days and MAD of 25.13 days which is the best performing
prediction model. We compare our results with the different ARIMA models
achieving an RMSE of 417.14 days and a MAD of 154.15 days. Prophet achieved
on median an RMSE of 443.75 days and a MAD of 119.0 days. The simple
moving average achieves an RMSE of 166.52 days and a MAD of 65.08 days.
Compared to the original SAFER analysis, our best results are still worse which
is expected in a larger, realistic data-set. This highlights the importance of our
extended evaluation here.

Predicting patch time-spans is not trivial, because the data has a high amount
of variation and ranges from a few days up to years. If the trend for all patch
intervals would be more stable, we believe that the auto-regressive model would
also estimate future patch-intervals more accurately. If the trend fluctuates heavily, Prophet has a higher residual error rate but a better adaptation to the
high variation in comparison to other evaluated prediction models. Moreover,

Prophet's multiplicative model better adapts to the heavily varying data in those forecasts but introduces a higher prediction error which significantly increases the RMSE in comparison to the more stable MAD. Based on the heavily varying data for device models and data sources, we achieved the most accurate prediction results when evaluating the best configuration per prediction model and data-set before forecasting.

To conclude, AR predicted the future patch intervals with an average error of 53.17 days in RMSE and 25.13 days in MAD. AR predicted the PT category for 318 out of 793 device models correctly. Due to the heavily varying data per device model and data-source, we do not suggest to choose one single prediction model and configuration for PT forecasting as this would worsen the prediction accuracy. For SAFER's PT, achieving the correct PT category based on predicted patch intervals for a device model is the most important task we want to solve. Thus, we discuss how accurate the predicted patch-intervals in days are in comparison to observed patch intervals of the test-set.

4.3 Prediction Accuracy

To show the accuracy of our SAFER version for each device model category like CCTV, we grouped the analyzed device models by their category in Table 2. The table shows that categorizing the PT (in *Fast*, *Medium* and *Slow*) correctly is not trivial due to less available and heavily fluctuating data, ranging from few days to multiple years for a single device model. We use the same patch trend categories (interval of vendor response) as Oser et al. [23]: *Fast* (0–22 days, the median time until an exploit exists), *Medium* (23–413 days, the median time until full disclosure) and *Slow* (≥414 days).

Table 2. Predicted PT accuracy per device category

Category	# D	Prophet	ARIMA	AR	SMA
CCTV	720	25.97%	20.41%	37.50%	30.41%
Streaming	63	33.33%	34.92%	46.03%	42.85%
Switch	4	0.0%	0.0%	0.0%	0.0%
Speaker	3	0.0%	0.0%	0.0%	0.0%
Controller	2	0.0%	50.0%	50.0%	0.0%
IP2Serial	1	100.0%	100.0%	100.0%	100.0%

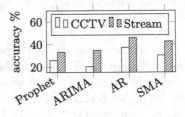

Table 2 shows that the predicted PT category does not often align with the observed PT category for all device models. Below, we discuss our findings for the prediction models to highlight which PT category was estimated and which category was correct. 318 out of 793 device models were predicted with the correct PT using AR. It wrongly estimated future PT for 52 device models (14% of all devices on which there was enough data to perform a prediction) with a too high/low PT category. We identified that 29 device models were estimated with a too high PT and 23 device models were estimated with a too low PT.

The lack of historic patch data prevented AR to predict the PT for 423 device models. We think that PT predictions will improve over time if vendors support devices longer, state when a patch was applied and when additional data-sources for SAFER are used to combine information. We also think that the prediction error can be decreased even further if vendors introduce regular patch-cycles resulting in less fluctuating patch intervals.

In this section, we presented a fine-grained evaluation of our enhanced SAFER patch trend calculation to estimate future patch intervals per device model and prediction model. The resulting patch trend, combining past and estimated future patch intervals, indicates in which time interval a vendor patches vulnerabilities. In standard risk assessments, the risk value is calculated by combining the incidence rate and the impact of the risk. For SAFER's device risk assessments, we consider the likelihood as the patch trend calculated above and the impact of the device risk as the vulnerabilities a device is vulnerable to. Hence, the following section focuses on the latter part: 1) considering a device's known vulnerabilities, and 2) estimating the severity of future vulnerabilities to 3) ultimately identify the most common vulnerability severity level per device model.

5   Predicting the Vulnerability Trend

In order to determine the likely severity levels of an IoT device's future vulnerabilities, we calculate the *vulnerability trend*. Similar to the previous section, we first introduce the data-set containing severity levels of past security vulnerabilities of our device models as it is provided by SAFER's firmware and release note analysis. Second, based on the device model's past vulnerabilities, we apply different prediction models to evaluate how well they are able to predict severity levels of future vulnerabilities. Third, we compare predicted future vulnerability severity levels with observed ones to identify our prediction accuracy.

5.1   Data-Set Observation

A histogram of all vulnerabilities for all analyzed device models without duplicates is shown in Fig. 2. Our enhanced version of SAFER identified that 293 out of *793* device models do not have registered vulnerabilities. This is either based on 1) vendors replacing firmware-contained software so that they do not contain vulnerable software or 2) SAFER did not find a vulnerability in analyzed firmware images.

Our data further contains 479 device models having between 1 and 2,064 registered vulnerabilities as shown in Fig. 2, arithmetic average being 549. Over all device models, the average amount of vulnerabilities is 341. Based on SAFER's retrieved data, the average vulnerability severity level over all models is 4.9 CVSS. The data is split into the first quartile ranging to 4.4 CVSS, the second quartile ranging to 4.9 CVSS and the third quartile ranging to 7.1 CVSS. Out of all retrieved vulnerabilities, the lowest CVSS severity level is 1.2 CVSS and

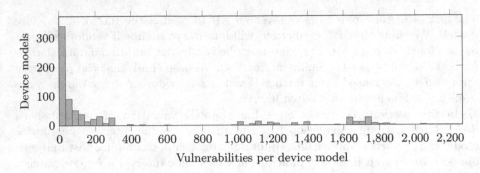

Fig. 2. All device models with more than one vulnerability

the highest CVSS is 10.0. To estimate future vulnerabilities for device models based on this data-set, we again analyzed different prediction models.

5.2   Prediction Model Evaluation

To estimate future vulnerability severity levels for the VT, we use the following prediction models; namely Facebook's Prophet, the auto-regressive integrated moving average (ARIMA) model, the auto-regressive model (AR) and a simple moving average (SMA) model. To evaluate the prediction models, we again split our data-set in a 66% training and 34% testing-set per device model. For the VT, the training and test-set contain the severity levels of all vulnerabilities per device model. As the VT data is based on vulnerability severity levels, the vertical axis is limited from the minimum 0.0 to the maximum 10.0 CVSS value. The forecast time interval (horizontal axis) is – analog to PT – set as the time interval of the test-set.

Our enhanced SAFER version was able to predict future vulnerability severity levels with Prophet for 793, with ARIMA for 770, with AR for 793 and with SMA for 793 out of 793 device models. Based on the correctly estimated device models and lowest prediction error, we consider AR as the best performing prediction model. SAFER requires a minimum amount of two past vulnerabilities to estimate a trend. If a device's firmware content has less than two vulnerabilities registered, the prediction models cannot predict future data accurately as there is no trend available. This happened for 394 device models for which we did not include a prediction error in Table 3. We argue that for those device models, the vendor actively takes care of replacing vulnerable software which reduces the VT data-set per device model and is represented in SAFER by indicating a *low* VT [23].

We identified the CVSS category of future vulnerabilities for all 793 device models correctly and achieved over all models an RMSE of 1.49 CVSS and a MAD of 0.98 CVSS using AR. This means that we can predict the severity for future vulnerabilities better than a simple moving average or Prophet which both achieve the same amount of correct device models but face a higher prediction error. Since the severity levels of the VT data range from 0.0 to 10.0 there is less fluctuating data. Thus, we use AR – being a linear model with less variation – to predict CVSS values from this data source.

We chose the best performing prediction model AR and discuss the predicted CVSS values hereafter.

5.3 Prediction Accuracy

In this section, we compare how accurate our SAFER version estimates the vulnerability trend in comparison to the observed test-data we use to verify the results.

Table 3. VT prediction accuracy overview

Version	Model	Corr. VT	Corr. #	RMSE	MAD
Our	Prophet	100.0%	793/793	2.08	1.40
Our	ARIMA	97.10%	770/793	1.94	1.04
Our	AR	100.0%	793/793	1.49	0.98
Our	SMA	100.0%	793/793	1.74	1.19
Orig	Prophet	100.0%	38/38	-	2.39
Orig	ARIMA	100.0%	38/38	-	1.31

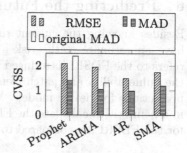

As Table 3 shows, three out of four prediction models estimated the VT correct for all devices. Only AR estimated the VT category for devices correct with the lowest RMSE of 1.49 CVSS and a MAD of 0.98 CVSS. Thus, our enhanced SAFER version reduces the prediction error in comparison to the original SAFER by 25.2%. To provide users of SAFER an intuitive understanding of the vulnerability trend, we categorize the most likely vulnerability severity by using CVSS version 2.0 into: *Low* (0.0–3.9), *Medium* (4.0–6.9) and *High* (7.0–10.0).

Table 4. Predicted VT accuracy per device category

Category	Devices	AR	ARIMA	SMA	Prophet
CCTV	720	100.0%	95.69%	100.0%	100.0%
Streaming	63	100.0%	100.0%	100.0%	100.0%
Switch	4	100.0%	100.0%	100.0%	100.0%
Speaker	3	100.0%	100.0%	100.0%	100.0%
Controller	2	100.0%	100.0%	100.0%	100.0%
IP2Serial	1	100.0%	100.0%	100.0%	100.0%

Table 4 shows that the VT category per device model was often predicted correctly by the prediction models. Apart from the ARIMA prediction model, all three others predicted the VT category correctly for 793 out of 793 device models. However, even though the prediction models achieved 100% for the VT category, they differ in their prediction errors measured in CVSS. We conclude that the prediction models of SAFER's enhanced version estimate the future vulnerability severity levels with 25.2% less error.

This section presented the evaluation of our enhanced SAFER version to estimate future vulnerability severity levels per device and prediction model. The resulting VT, combining past and estimated future vulnerability severity levels, indicates the most likely vulnerability severity per device model. Combining 1) the most likely time a vendor requires to issue a firmware update (by PT) with 2) the most likely vulnerability severity level per device model (by VT) results in the Future Device Security Risk Indicator (FDSRI) described below.

6   Predicting the Future Device Security Risk Indicator

Besides knowing the current risk, it is also relevant for IoT device owners to have an estimate of the risk a device might pose in the future. This section refers to the FDSRI introduced in SAFER's original work [23] which is based on the vulnerability and patch trend, both containing past observations and future predictions for device models. We use the patch trend and vulnerability trend in a heuristic to calculate the FDSRI. Then, we discuss how accurate the results for the FDSRI are compared to real, observed data.

6.1   Definition of the Future Device Security Risk Indicator

Table 5. Future Device Security Indicators

Vulnerability Trend	Patch Trend		
	Fast	Medium	Slow
Low	Low	Low	Medium
Medium	Low	Medium	High
High	Medium	High	Critical

To determine the FDSRI per device model, SAFER needs to combine the device model's previously calculated PT and VT. We use the risk matrix (Table 5) introduced in SAFER's original work [23] to combine the PT and VT in a heuristic leading to the FDSRI. We evaluated our approach by first calculating the VT and PT on observed data including the FDSRI. Afterwards, SAFER predicted the patch intervals and future vulnerability severity levels for the test-set which we used in previous sections to calculate the PT and VT on. Ultimately, we calculated the FDSRI using both trend (PT and VT) predictions and compare it with the FDSRI derived from real observations. This allows us to verify the accuracy of SAFER's estimated FDSRI.

We highlight that our enhanced SAFER version estimates the correct PT for 40.10% and the correct VT for 100% of 793 device models. We achieve this by individually parameterizing and combining different prediction models to best fit the patch and vulnerability data.

6.2 Classifying the Future Device Security Risk Indicator

Table 6. FDSRI accuracy comparison.

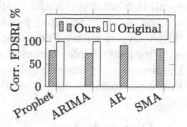

Version	PT/VT Model	Corr. FDSRI	Corr. #
Our	Prophet	79.70%	632/793
Our	ARIMA	73.52%	583/793
Our	AR	91.30%	724/793
Our	SMA	84.49%	670/793
Original	Prophet	100%	38/38
Original	ARIMA	100%	38/38

Table 6 shows how many future device security risk indicators (FDSRI) were identified correctly. In previous sections for VT and PT, we identified that the AR prediction model performed best for the data-source which Table 6 shows by achieving a correct FDSRI for 91.30%, i.e. 724 of 793 device models.

Analogous to VT and PT separately, the original SAFER evaluation with a small sample of devices had an unrealistically high accuracy. Our evaluation therefore provides a realistic prediction performance.

Table 7. Correctly estimated FDSRI per device category

Category	Correct FDSRI	Devices per Category
CCTV	90.83%	720
Streaming	95.24%	63
Switch	100.0%	4
Speaker	100.0%	3
Controller	100.0%	2
IP2Serial	100.0%	1

When looking at the device model categories in Table 7, our SAFER version estimated the FDSRIs for the largest device category (CCTV) with 721 device models achieving a correct FDSRI for 90.84% and the second largest device category (Streaming) achieved a correct FDSRI for 95.16%. 41 out of 793 device models are estimated to have a higher FDSRI than we observed in real data. This splits into two device models where SAFER predicted the FDSRI to be *medium* but was observed *low*, and 39 device models where SAFER predicted a *high* FDSRI but was observed *medium*. On the contrary, for 53 device models SAFER estimated a too low FDSRI. Those divide in 16 device models with estimated *medium* and observed *high* FDSRI, and 37 device models with estimated *low* and observed *medium* FDSRI. The reason being that the PT was previously predicted too high for 29 device models (*medium* instead of *slow*) and for 23 device models, SAFER predicted a too low PT (*slow* instead of *medium*).

We showed that our SAFER version – implementing the above mentioned mechanisms to predict PT, VT and FDSRI – is able to indicate the future device security risk indicator for *91.30% of 793* device models correctly. Moreover, we recall that after the initial configuration, SAFER is able to perform device risk scoring in an automated fashion. This enables SAFER to inform embedded-device owners with different security skill-sets about the estimated future device security risks, display technical evidence, e.g. discovered software libraries within the firmware, to understand the risks and to support SAFER's users in making informed decisions about possible mitigations to secure the connected network.

7   Discussion

We point out that we split the data into a training and test-set for predictions of PT and VT. Using this split, we verified that all predictions were compared to observed data. This enabled us to make statements on how accurate the prediction mechanisms perform compared to real, observed data.

7.1   Limitations

We assume that publicly registered vulnerabilities for device models and third-party software are verified by other parties prior to registering them at CVE numbering authorities. However, SAFER cannot identify if the device uses, e.g., a vulnerable function in its operation. SAFER estimates the device risks based on publicly known vulnerabilities and patch intervals. This does not include unpublished vulnerabilities (zero-days) from, e.g., nation states and underground forums. Thus, SAFER's calculated risk indicator needs to be considered as a risk approximation based on public vulnerability information. SAFER considers the "created" date of a CVE as the date the CVE was publicly registered. SAFER cannot identify if at this date the CVE contained vulnerability information or temporary placeholder information. We assume that when a CVE gets registered for a product, the vendor knows about the vulnerability and is already able to work on patches. The extended data-set from this work contains data from Axis devices, because we considered license statements and release note files as data-sources from this vendor. A wider range of vendors would result in a more realisitc data-set and alleviate possible a possible bias specific to Axis. However, with SAFER, one can create the metrics we discussed for Axis for various vendors, e.g., by uploading firmwares to SAFER's firmware analysis component.

7.2   Comparing with Related Work

The majority of vulnerability prediction approaches require the public availability of source code for all software versions. Massacci et al. [21], for example, investigated Firefox and compared 18 different vulnerability approaches of related work with their own approach, whereas the 12 best performing ones focus on vulnerability prediction. Other vulnerability prediction works require specific amount of past software patches in the source-code as, e.g., [9] to train a machine model on. Other works are limited to specific programming languages they can parse and process as, e.g., [5]. The last group does not only require source code availability, but also a large code-base as, e.g., [12] to perform predictions on. Unlike related work [6,20,25,29,37,38] finding vulnerabilities in, e.g., source-code, SAFER uses public sources to fetch known vulnerabilities only. Though, related work [8,15,35,36] with the same approach achieved accurate results. However, we state that by using SAFER's approach to retrieve vulnerabilities, SAFER is not limited in detecting vulnerabilities for, e.g., specific programming languages, but even supports closed-source software and does not require a large source-code base. SAFER's risk scoring component uses the CVSS standard (in comparison to modified CVSS versions like related work [10,26,33,34]) to generate universal and comparable metrics. This is a relevant fact to make device comparisons for different device categories possible.

8   Summary and Future Work

This paper presented an in-depth evaluation of the FDSRI with 793 device models in comparison to its original work [23] with 38 device models and showed that historic vulnerability and patch data can be used to estimate future risks. We used SAFER's risk *prediction & scoring* component, which aims to automate future security risk assessments by calculating the future device security risk indicator (FDSRI). This indicator identifies how problematic a device might become in the future even if the currently used firmware is considered secure.

With our enhanced SAFER version, we decreased the vulnerability trend's prediction error by 25.2% and slightly increased the patch trend's prediction error by merely 4.1%. We highlight that in comparison to SAFER's original work [23] with 38 device models, we analyzed 793 IoT devices which represents 20.86 times more device models. Hence, our evaluation provides a clear indication of the feasibility but also scalability of SAFER.

SAFER is a suitable tool to establish IoT security awareness in large-scale networks and enables highly automated risk assessments for IoT devices estimating its current as well as future risk. By using our enhanced SAFER version, we correctly estimate the FDSRI for 91.30% of 793 devices.

8.1   Future Work

The risk *prediction & scoring* component could be further enhanced by implementing detection methods for vulnerability chaining, e.g., by using the Common Weakness Enumeration (CWE).

Assuming that SAFER will include more data-sources in the future, we consider evaluating Long Short-Term Memory (LSTM) and further neural network based prediction models for future work which require significantly more data to compute predictions on.

Acknowledgement. This work was partially funded by the Sapere Aude: DFF-Starting Grant number 0165-00079B "Foundations of Privacy Preserving and Accountable Decentralized Protocols" and by the Wolfgang Gentner Programme of the German Federal Ministry of Education and Research (grant no. 05E15CHA).

References

1. Agarwal, S., Oser, P., Lueders, S.: Detecting IoT devices and how they put large heterogeneous networks at security risk. Sensors **19**(19), 4107 (2019)
2. Bahizad, S.: Risks of increase in the IoT devices. In: 2020 7th IEEE International Conference on Cyber Security and Cloud Computing (CSCloud)/2020 6th IEEE International Conference on Edge Computing and Scalable Cloud (EdgeCom), pp. 178–181. IEEE (2020)
3. Box, G.E., Jenkins, G.M., Reinsel, G.C.: Time Series Analysis: Forecasting and Control, vol. 734. Wiley, Hoboken (2011)
4. Chakraborty, S., Krishna, R., Ding, Y., Ray, B.: Deep learning based vulnerability detection: are we there yet. IEEE Trans. Softw. Eng. (2021)
5. Dam, H.K., Tran, T., Pham, T., Ng, S.W., Grundy, J., Ghose, A.: Automatic feature learning for vulnerability prediction. arXiv preprint arXiv:1708.02368 (2017)
6. Duan, R., et al.: Automating patching of vulnerable open-source software versions in application binaries. In: NDSS (2019)
7. Duan, X., et al.: Automated security assessment for the internet of things. arXiv preprint arXiv:2109.04029 (2021)
8. Edkrantz, M., Truvé, S., Said, A.: Predicting vulnerability exploits in the wild. In: 2015 IEEE 2nd International Conference on Cyber Security and Cloud Computing, pp. 513–514. IEEE (2015)
9. Garg, A., Degiovanni, R., Jimenez, M., Cordy, M., Papadakis, M., Traon, Y.L.: Learning to predict vulnerabilities from vulnerability-fixes: a machine translation approach. arXiv preprint arXiv:2012.11701 (2020)
10. Guillen, O.M., Brederlow, R., Ledwa, R., Sigl, G.: Risk management in embedded devices using metering applications as example. In: Proceedings of the 9th Workshop on Embedded Systems Security, pp. 1–9 (2014)
11. Howell, D.C.: Median absolute deviation. Encyclopedia of Statistics in Behavioral Science (2005)
12. Jimenez, M., Papadakis, M., Le Traon, Y.: Vulnerability prediction models: a case study on the Linux kernel. In: 2016 IEEE 16th International Working Conference on Source Code Analysis and Manipulation (SCAM), pp. 1–10. IEEE (2016)
13. Johnson, P., Lagerström, R., Ekstedt, M., Franke, U.: Can the common vulnerability scoring system be trusted? A Bayesian analysis. IEEE Trans. Dependable Secure Comput. **15**(6), 1002–1015 (2016)
14. Jøsang, A.: Subjective Logic: A Formalism for Reasoning Under Uncertainty. Artificial Intelligence: Foundations, Theory and Algorithms. Springer, Heidelberg (2016). https://doi.org/10.1007/978-3-319-42337-1
15. Kudjo, P.K., Chen, J., Mensah, S., Amankwah, R., Kudjo, C.: The effect of bellwether analysis on software vulnerability severity prediction models. Softw. Qual. J. **28**(4), 1413–1446 (2020)
16. Le, N.T., Hoang, D.B.: Security threat probability computation using Markov chain and common vulnerability scoring system. In: 2018 28th International Telecommunication Networks and Applications Conference (ITNAC), pp. 1–6. IEEE (2018)

17. Li, Q., Tan, D., Ge, X., Wang, H., Li, Z., Liu, J.: Understanding security risks of embedded devices through fine-grained firmware fingerprinting. IEEE Trans. Dependable Secure Comput. **19**, 4099–4112 (2021)
18. Li, Z., Zou, D., Xu, S., Jin, H., Zhu, Y., Chen, Z.: SySeVR: a framework for using deep learning to detect software vulnerabilities. IEEE Trans. Dependable Secure Comput. **19**, 2244–2258 (2021)
19. Li, Z., et al.: VulDeePecker: a deep learning-based system for vulnerability detection. arXiv preprint arXiv:1801.01681 (2018)
20. Liu, B., Shi, L., Cai, Z., Li, M.: Software vulnerability discovery techniques: a survey. In: 2012 Fourth International Conference on Multimedia Information Networking and Security, pp. 152–156. IEEE (2012)
21. Massacci, F., Nguyen, V.H.: Which is the right source for vulnerability studies? An empirical analysis on Mozilla Firefox. In: Proceedings of the 6th International Workshop on Security Measurements and Metrics, pp. 1–8 (2010)
22. Oser, P., et al.: Safer: development and evaluation of an IoT device risk assessment framework in a multinational organization. Proc. ACM Interact. Mob. Wearable Ubiquit. Technol. **4**(3), 1–22 (2020)
23. Oser, P., van der Heijden, R.W., Lüders, S., Kargl, F.: Risk prediction of IoT devices based on vulnerability analysis. ACM Trans. Priv. Secur. **25**(2), 1–36 (2022)
24. Oser, P., Kargl, F., Lüders, S.: Identifying devices of the internet of things using machine learning on clock characteristics. In: Wang, G., Chen, J., Yang, L.T. (eds.) SpaCCS 2018. LNCS, vol. 11342, pp. 417–427. Springer, Cham (2018). https://doi.org/10.1007/978-3-030-05345-1_36
25. Perl, H., et al.: VCCFinder: finding potential vulnerabilities in open-source projects to assist code audits. In: Proceedings of the 22nd ACM SIGSAC Conference on Computer and Communications Security, pp. 426–437 (2015)
26. Qu, Y., Chan, P.: Assessing vulnerabilities in Bluetooth low energy (BLE) wireless network based IoT systems. In: 2016 IEEE 2nd International Conference on Big Data Security on Cloud (BigDataSecurity), IEEE International Conference on High Performance and Smart Computing (HPSC), and IEEE International Conference on Intelligent Data and Security (IDS), pp. 42–48. IEEE (2016)
27. Rodríguez, E., Noroozian, A., van Eeten, M., Gañán, C.: Superspreaders: quantifying the role of IoT manufacturers in device infections (2021)
28. Russell, R., et al.: Automated vulnerability detection in source code using deep representation learning. In: 2018 17th IEEE International Conference on Machine Learning and Applications (ICMLA), pp. 757–762. IEEE (2018)
29. Shin, Y., Meneely, A., Williams, L., Osborne, J.A.: Evaluating complexity, code churn, and developer activity metrics as indicators of software vulnerabilities. IEEE Trans. Softw. Eng. **37**(6), 772–787 (2010)
30. Shivraj, V., Rajan, M., Balamuralidhar, P.: A graph theory based generic risk assessment framework for internet of things (IoT). In: 2017 IEEE International Conference on Advanced Networks and Telecommunications Systems (ANTS), pp. 1–6. IEEE (2017)
31. Taylor, S.J., Letham, B.: Forecasting at scale. Am. Stat. **72**(1), 37–45 (2018)
32. Vilches, V.M., et al.: Towards an open standard for assessing the severity of robot security vulnerabilities, the robot vulnerability scoring system (RVSS). arXiv preprint arXiv:1807.10357 (2018)
33. Wang, H., Chen, Z., Zhao, J., Di, X., Liu, D.: A vulnerability assessment method in industrial internet of things based on attack graph and maximum flow. IEEE Access **6**, 8599–8609 (2018)

34. Wang, R., Gao, L., Sun, Q., Sun, D.: An improved CVSS-based vulnerability scoring mechanism. In: 2011 Third International Conference on Multimedia Information Networking and Security, pp. 352–355. IEEE (2011)
35. Williams, M.A., Barranco, R.C., Naim, S.M., Dey, S., Hossain, M.S., Akbar, M.: A vulnerability analysis and prediction framework. Comput. Secur. **92**, 101751 (2020)
36. Wu, S., Wang, C., Zeng, J., Wu, C.: Vulnerability time series prediction based on multivariable LSTM. In: 2020 IEEE 14th International Conference on Anticounterfeiting, Security, and Identification (ASID), pp. 185–190. IEEE (2020)
37. Xiao, Y., et al.: MVP: detecting vulnerabilities using patch-enhanced vulnerability signatures. In: 29th USENIX Security Symposium (USENIX Security 2020), pp. 1165–1182 (2020)
38. Xu, Z., Chen, B., Chandramohan, M., Liu, Y., Song, F.: Spain: security patch analysis for binaries towards understanding the pain and pills. In: 2017 IEEE/ACM 39th International Conference on Software Engineering (ICSE), pp. 462–472. IEEE (2017)
39. Zhou, Y., Liu, S., Siow, J., Du, X., Liu, Y.: Devign: effective vulnerability identification by learning comprehensive program semantics via graph neural networks. arXiv preprint arXiv:1909.03496 (2019)

Server-Supported Decryption for Mobile Devices

Johanna Maria Kirss[1,2]($\boxtimes$), Peeter Laud[1]($\boxtimes$), Nikita Snetkov[1,3],
and Jelizaveta Vakarjuk[1]

[1] Cybernetica AS, Tallinn, Estonia
{johanna.kirss,peeter.laud,nikita.snetkov,
jelizaveta.vakarjuk}@cyber.ee
[2] University of Tartu, Tartu, Estonia
[3] Tallinn University of Technology, Tallinn, Estonia

Abstract. We propose a threshold encryption scheme with two-party
decryption, where one of the keyshares may be stored and used in a
device that is able to provide only weak security for it. We state the
security properties the scheme needs to have to support such use-cases,
and construct a scheme with these properties. Our construction is based
on the ElGamal cryptosystem, with additional zero-knowledge proofs
that can provide IND-CCA security, and resistance to offline guessing
attacks.

Keywords: Threshold encryption schemes · Offline guessing attacks

1 Introduction

Considering recent legistlative initiatives [14], we may be soon storing many verifiable credentials about our sensitive attributes in our smartphones, supported by electronic wallet applications, streamlining the procurement and presentation of these credentials. By themselves, smartphones cannot provide sufficient confidentiality for these credentials. Rather, we expect to store them in some encrypted form, decrypted only while they are in use. The decryption keys are stored inside a Secure Element [22], a tamper-resistant piece of hardware contained in the phone. Hence we have to trust the producers of Secure Elements, and abstain from the use of e-Wallets (with credentials containing sensitive information) on phones without Secure Elements. This is not a desirable situation. We would like to replace trusted hardware with something that has weaker trust requirements, e.g. threshold cryptography.

Common constructions of primitives of threshold cryptography and their security definitions are difficult to map to a setting where some keyshares are stored and operations performed on platforms with weak security protections. We have proposed a server-supported signature scheme [7], where the signing key was shared between a phone and a server, and the keyshare in the phone was protected only by symmetrically encrypting it with a key with very low entropy (derived from a PIN that the user can remember). The security of our scheme

G. Lenzini and W. Meng (Eds.): STM 2022, LNCS 13867, pp. 71–81, 2023.
https://doi.org/10.1007/978-3-031-29504-1_4

was based on the infeasibility of offline guessing attacks by someone who has obtained the encrypted keyshare of the phone, and by the ability of the server to recognize online guessing attacks. The latter property allows the server to count wrong guesses, and a clone detection mechanism [19] allows to reset the counter.

In this paper, we propose an encryption scheme with similar properties, i.e. it has distributed decryption, where offline guesses by someone masquerading as the phone are impossible, and wrong guesses made online are detected by the server. We want the phone to initiate the decryption, and the server to learn nothing about the decrypted plaintext. We give a formalization of these properties. Combined with the clone detection mechanism, our scheme could be used as an alternative to Secure Elements, *at least* when the requirement for online connectivity during decryption is acceptable.

Related Work. Our encryption scheme is motivated by a set of requirements that have previously not been tried to address together. They have been considered in the context of server-supported signature schemes [7], where we attempt to avoid *offline guessing attacks* [2] and detect *online guessing attacks* [12]. This is also the case for scheme [8], where the server supports the functionality of a secure construction. It is in contrast to the schemes where a server is employed to reduce the client's workload in performing computationally expensive operations [1,4]. It is also in contrast to *key-insulated encryption* [13], where a mostly offline server is used to reduce the impact of repeatedly breaking a weakly secure device.

Our scheme builds upon threshold encryption schemes. Threshold cryptography has a long history, starting from [11], where a method for threshold creation of RSA signatures was proposed. IND-CCA secure encryption schemes with threshold decryption [21] were proposed shortly after IND-CCA secure asymmetric encryption schemes [10]. At present, threshold cryptography is a mature field, discussed in textbooks [5] and subject to standardization activities [6].

2 Desired Properties of Distributed Decryption

In this paper, we consider asymmetric *key encapsulation* [18] schemes, where the decapsulation functionality is distributed between two parties—the *client*, and the *server*. The roles of these parties are not identical, and the desired security properties for each of them are different.

An encapsulation scheme with client-server decryption consists of the following sets, algorithms, and protocols, parameterized with the security parameter λ and other public parameters (e.g. the definition of the used cyclic groups):

- Sets of shared secrets SS, ciphertexts CT, public keys PK, client's private keys $\mathsf{SK_C}$, and server's private keys $\mathsf{SK_S}$.
- Key-generation protocol $\langle \mathcal{KG}_\mathsf{C} | \mathcal{KG}_\mathsf{S} \rangle$, run by both parties. It returns $(\mathsf{sk}_1, \mathsf{pk}) \in \mathsf{SK_C} \times \mathsf{PK}$ to the client, and $(\mathsf{sk}_2, \mathsf{pk}) \in \mathsf{SK_S} \times \mathsf{PK}$ to the server.
- Encapsulation algorithm $\mathcal{E}nc$. It takes as input a public key $\mathsf{pk} \in \mathsf{PK}$, and returns a shared secret $k \in \mathsf{SS}$ and a ciphertext $c \in \mathsf{CT}$.

Experiment IND-CCA-S$^{\mathcal{A}}$	Experiment IND-CCA-C$^{\mathcal{A}}$
$\langle(\mathsf{sk}_1,\mathsf{pk}),(\mathsf{sk}_2,\mathsf{state})\rangle \leftarrow_\$ \langle\mathcal{KG}_C|\mathcal{A}()\rangle$	$\langle(\mathsf{sk}_1,\mathsf{state}),(\mathsf{sk}_2,\mathsf{pk})\rangle \leftarrow_\$ \langle\mathcal{A}()|\mathcal{KG}_S\rangle$
$(k_0,c) \leftarrow_\$ \mathcal{E}nc(\mathsf{pk})$	$(k_0,c) \leftarrow_\$ \mathcal{E}nc(\mathsf{pk})$
$k_1 \leftarrow_\$ \mathsf{SS},\ b \leftarrow_\$ \{0,1\}$	$k_1 \leftarrow_\$ \mathsf{SS},\ b \leftarrow_\$ \{0,1\}$
$\mathcal{O}_1(\cdot) \leftarrow \mathcal{D}ec(\cdot,\mathsf{sk}_1,\mathsf{sk}_2,\mathsf{pk})$	$\mathcal{O}_1(\cdot) \leftarrow \mathcal{D}ec(\cdot,\mathsf{sk}_1,\mathsf{sk}_2,\mathsf{pk})$
$\mathcal{O}_2(\cdot) \leftarrow \mathcal{DC}_C(\cdot,\mathsf{sk}_1,\mathsf{pk})$	$\mathcal{O}_2(\cdot) \leftarrow \mathcal{DC}_S(\cdot,\mathsf{sk}_2,\mathsf{pk})$
$b^* \leftarrow_\$ \mathcal{A}^{\mathsf{EXCL}_c[\mathcal{O}_1],\mathsf{EXCL}_c[\mathcal{O}_2]}(\mathsf{state},c,k_b)$	$b^* \leftarrow_\$ \mathcal{A}^{\mathsf{EXCL}_c[\mathcal{O}_1],\mathsf{EXCL}_c[\mathcal{O}_2]}(\mathsf{state},c,k_b)$
return $b = b^*$	**return** $b = b^*$

Fig. 1. Security against chosen-ciphertext attacks

- Decapsulation protocol $\langle\mathcal{DC}_C|\mathcal{DC}_S\rangle$, run by the client and the server. Client's inputs are $c \in \mathsf{CT}$, $\mathsf{sk}_1 \in \mathsf{SK}_C$, and $\mathsf{pk} \in \mathsf{PK}$. Server's inputs are $c \in \mathsf{CT}$, $\mathsf{sk}_2 \in \mathsf{SK}_S$ and $\mathsf{pk} \in \mathsf{PK}$. The protocol returns either $k \in \mathsf{SS}$ or the failure notice $\perp$ to the client. It returns success/failure notice $\top/\perp$ to the server.

We also define the *decapsulation algorithm* $\mathcal{D}ec$, that on inputs $c,\mathsf{sk}_1,\mathsf{sk}_2,\mathsf{pk}$ invokes $\langle\mathcal{DC}_C(c,\mathsf{sk}_1,\mathsf{pk})|\mathcal{DC}_S(\mathsf{sk}_2,\mathsf{pk})\rangle$ and returns client's output.

In the following, we write $x_1,\ldots,x_n \leftarrow_\$ X$ to denote that values $x_1,\ldots,x_n$ are uniformly, independently sampled from a set X. We also write $x \leftarrow_\$ \mathsf{X}(\ldots)$ to denote that x is returned by a stochastic computation X. Given an oracle $\mathcal{O}(\cdot)$ and a value c, we let $\mathsf{EXCL}_c[\mathcal{O}]$ denote an oracle that on input c^* returns $\perp$ if $c^* = c$, and $\mathcal{O}(c^*)$ otherwise. A protocol party executed as an oracle gives the adversary the messages this party produces.

Definition 1 (Correctness). *A encapsulation scheme is* correct, *if*

$$\Pr\left[k' = k \wedge r = \top \;\middle|\; \begin{array}{l} \langle(\mathsf{sk}_1,\mathsf{pk}),(\mathsf{sk}_2,\mathsf{pk})\rangle \leftarrow_\$ \langle\mathcal{KG}_C|\mathcal{KG}_S\rangle, (k,c) \leftarrow \mathcal{E}nc(\mathsf{pk}), \\ \langle k',r\rangle \leftarrow \langle\mathcal{DC}_C(c,\mathsf{sk}_1,\mathsf{pk})|\mathcal{DC}_S(c,\mathsf{sk}_2,\mathsf{pk})\rangle \end{array}\right] \approx 1.$$

The confidentiality properties of the encapsulation scheme are defined in the usual manner. The definitions refer to the experiments in Fig. 1 that follow general definitions of IND-CCA for threshold encryption schemes.

Definition 2 (IND-CCA against server/client). *The encapsulation scheme provides indistinguishability against the chosen-ciphertext attacks by the server [resp. client], if the experiment IND-CCA-S$^{\mathcal{A}}$ [resp. IND-CCA-C$^{\mathcal{A}}$] is successful with probability at most negligibly larger than $1/2$ for all efficient adversaries $\mathcal{A}$.*

The impossibility of offline guessing and detectability of online guessing is defined below, using the experiment defined in Fig. 2. Here shuffle returns a list that is a random permutation of its arguments. The list SK_1 corresponds to the list of candidate private keys that an intruder may obtain after extracting the weakly encrypted (e.g. the encryption key has been derived from a PIN) private

Experiment OG-CCA-C$_{T,L}^{\mathcal{A}}$

$\langle(\mathsf{sk}_1^{(1)}, \mathsf{pk}), (\mathsf{sk}_2, \mathsf{pk})\rangle \leftarrow_\$ \langle\mathcal{KG}_\mathrm{C}|\mathcal{KG}_\mathrm{S}\rangle$

$\mathsf{sk}_1^{(2)}, \ldots, \mathsf{sk}_1^{(L)} \leftarrow_\$ \mathsf{SK}_\mathrm{C},\ SK_1 \leftarrow_\$ \mathsf{shuffle}(\mathsf{sk}_1^{(1)}, \ldots, \mathsf{sk}_1^{(L)})$

$(k_0, c) \leftarrow_\$ \mathcal{E}nc(\mathsf{pk}),\ k_1 \leftarrow_\$ \mathsf{SS},\ b \leftarrow_\$ \{0,1\},\ t \leftarrow 0$

$\mathcal{O}_2(\cdot) \leftarrow \big\{\ t \leftarrow t+1;\ r \leftarrow \mathcal{DC}_\mathrm{S}(\cdot, \mathsf{sk}_2, \mathsf{pk});\ \mathbf{if}(r = \top)\ \mathbf{then}\ t \leftarrow t-1\ \big\}$

$b^* \leftarrow_\$ \mathcal{A}^{\mathsf{EXCL}_c[\mathcal{D}ec(\cdot, \mathsf{sk}_1^{(1)}, \mathsf{sk}_2, \mathsf{pk})], \mathcal{O}_2(\cdot)}(\mathsf{pk}, c, k_b, SK_1)$

return $(b = b^*)$ and $t \leq T$

Fig. 2. Security against offline and online guessing

key from the smartphone, and trying to decrypt it with all possible values of the key. We see that the adversary may start sessions of the server, and may even submit it the challenge ciphertext, but no more than T sessions may finish with $\bot$ (or not finish at all).

Definition 3 (No guessing by client). *The encapsulation scheme provides offline guessing security against chosen-ciphertext attacks by the client, if the experiment OG-CCA-C$_{T,L}^{\mathcal{A}}$ is successful with probability at most negligibly larger than $1/2 + T/L$ for all efficient adversaries $\mathcal{A}$, and numbers T, L.*

Finally, we ask for the integrity of shared secrets, i.e. the client would not accept a secret k' different from the one output by the encapsulation algorithm.

Definition 4 (Integrity for client). *The encapsulation scheme provides key integrity for the client, if for all efficient adversaries $\mathcal{A}$,*

$$\Pr\left[k' \in \{k, \bot\}\ \middle|\ \begin{array}{l} \langle(\mathsf{sk}_1, \mathsf{pk}), \mathsf{state}\rangle \leftarrow \langle\mathcal{KG}_\mathrm{C}|\mathcal{A}()\rangle, (k, c) \leftarrow \mathcal{E}nc(\mathsf{pk}), \\ \langle k', {\_}\rangle \leftarrow \langle\mathcal{DC}_\mathrm{C}(c, \mathsf{sk}_1, \mathsf{pk})|\mathcal{A}(\mathsf{state})\rangle \end{array}\right] \approx 1.$$

3 Building Blocks

Let $\mathbb{G}$ be a cyclic group of size p, with generator g. The *discrete logarithm problem* is to find $n \in \mathbb{Z}_p$, such that $g^n = h$, for a value $h \leftarrow_\$ \mathbb{G}$. The *decisional Diffie-Hellman (DDH) problem* is to distinguish tuples of the form (g, g^x, g^y, g^{xy}) (called *Diffie-Hellman tuples*) from tuples of the form (g, g^x, g^y, g^z) for $x, y, z \leftarrow_\$ \mathbb{Z}_p$. A problem is *hard* if all efficient algorithms have at most negligible advantage (over a trivial algorithm) of solving it.

Our schemes build on top of the ElGamal KEM, the IND-CPA security of which is equivalent to the hardness of DDH in the used group $\mathbb{G}$. In this KEM, private key is a random $\mathsf{sk} \in \mathbb{Z}_p$, while public key is $\mathsf{pk} = g^{\mathsf{sk}}$. The encapsulation algorithm generates $r \leftarrow_\$ \mathbb{Z}_p$, and outputs the shared secret $ss \leftarrow \mathsf{pk}^r$ and ciphertext $c = g^r$. The decapsulation algorithm computes $ss = c^{\mathsf{sk}}$. In *hashed ElGamal*, the shared secret is $H'(\mathsf{pk}^r)$ for some hash function H' that we model as a random oracle. Note that an input to a random oracle can be anything encodable as a bitstring.

A *DDH proof* $\pi = (\alpha, \alpha', \gamma) \leftarrow_\$ \mathsf{DHP}^H[r|_{g,h}^{u,v}|ctx]$ [9] is a non-interactive [16] zero-knowledge (NIZK) proof that $\log_g u = \log_h v$, given in *context ctx*, where $r \in \mathbb{Z}_p$ is the discrete logarithm and H is a hash function, modeled as a random oracle. It is given by $s \leftarrow_\$ \mathbb{Z}_p$, $\alpha \leftarrow g^s$, $\alpha' \leftarrow h^s$, $\beta \leftarrow H(g, h, u, v, \alpha, \alpha', ctx) \in \mathbb{Z}_p$, and $\gamma \leftarrow s + r \cdot \beta$. The checking procedure $\mathsf{ChP}^H[\pi|_{g,h}^{u,v}|ctx]$ recomputes β, and checks that $g^\gamma = \alpha \cdot u^\beta$ and $h^\gamma = \alpha' \cdot v^\beta$.

In our schemes, similarly to [21], DDH proofs are often used to give *simulatable proofs of knowledge of exponent* $\pi' \leftarrow_\$ \mathsf{KnE}^{H,\tilde{H}}[r|_g^u|ctx]$. These prove that someone knows the value $r = \log_g u$. Additionally, they allow the simulator to raise a value (an element of $\mathbb{G}$) of its choice to the power of r; the simulator has to choose that value at the time the adversary computes the proof. The construction makes use of two hash functions, both modeled as random oracles, where H returns elements of $\mathbb{Z}_p$ and $\tilde{H}$ returns elements of $\mathbb{G}$. It is given by first computing $h \leftarrow \tilde{H}(g, u, ctx)$ and $v \leftarrow h^r$. The proof is $\pi' = (\pi, v)$, where $\pi \leftarrow \mathsf{DHP}^H[r|_{g,h}^{u,v}|ctx]$. The checking procedure $\mathsf{ChE}^{H,\tilde{H}}[\pi'|_g^u|ctx]$ recomputes v and checks the DDH proof. During simulation, if the simulator wants to obtain z^r, it will generate $t \leftarrow_\$ \mathbb{Z}_p$, and program $\tilde{H}$ to return $z^{1/t}$ when the adversary queries it with g, u, ctx. Then $z^r = v^t$.

4   The Encryption Scheme

Secret-sharing the private key will straightforwardly thresholdize the ElGamal KEM [15]. IND-CCA may be achieved by adding the *non-interactive zero-knowledge proof of knowledge* (NIZKPoK) of r to the ciphertext. For non-threshold systems, this may be a designated-verifier (DV) NIZKPoK [10], aimed towards the receiver. For general case, Schnorr's proofs for discrete logarithm [20], made non-interactive through the Fiat-Shamir transform [16] using a random oracle (i.e. Schnorr signatures, using r as the signing key), are typically used to show the knowledge of r, but it is unknown how to combine them with ElGamal KEM in a way that allows IND-CCA to be derived only from the hardness of the DDH problem [3]. The TDH2 (threshold) cryptosystem [21] overcomes this by changing how the random oracle is used by the Fiat-Shamir transform, making certain additional computations possible in the simulation. The scheme NPS that we present here is quite similar to TDH2. Interestingly, only small changes are needed to make it secure against guessing attacks (Definition 3).

Let $\mathbb{G}$ be a cyclic group of size p, with generator g, with hard DDH problem. Let H_1, H_2, H_3 be hash functions outputting elements of $\mathbb{Z}_p$, and $\tilde{H}_1, \tilde{H}_2$ be hash functions outputting elements of $\mathbb{G}$, all modeled as random oracles. We put $\mathsf{NPS.SS} = \mathsf{NPS.PK} = \mathbb{G}$, $\mathsf{NPS.SK_C} = \mathbb{Z}_p$, and $\mathsf{NPS.SK_S} = \mathbb{Z}_p \times \mathbb{G}^2$. The set $\mathsf{NPS.CT}$ is given together with the algorithm $\mathsf{NPS}.\mathcal{E}nc$ and protocol $\mathsf{NPS}.\mathcal{DC}$ in Fig. 3. In the key-generation protocol, the client generates $\mathsf{sk}_1 \leftarrow_\$ \mathbb{Z}_p$, and the server generates $\mathsf{sk}_2 \leftarrow_\$ \mathbb{Z}_p$. They compute $\mathsf{pk}_i \leftarrow g^{\mathsf{sk}_i}$, fairly exchange the values pk_i with each other (using some *trapdoor commitment* scheme [17]), and define $\mathsf{pk} \leftarrow \mathsf{pk}_1 \cdot \mathsf{pk}_2$. The server stores $\mathsf{pk}_1, \mathsf{pk}_2$ together with the private exponent sk_2.

NPS.$\mathcal{E}nc$(pk)	NPS.$\mathcal{DC}_C((u,\pi),\mathsf{sk}_1,\mathsf{pk})$	NPS.$\mathcal{DC}_S((u,\pi),\mathsf{sk}_2,\mathsf{pk}_1,\mathsf{pk}_2,\mathsf{pk})$
$r \leftarrow_s \mathbb{Z}_p$	$\mathsf{ChE}^{H_1,\tilde{H}_1}[\pi\,|\,{}^{u}_{g}]$	$P_1 \longrightarrow: u, \pi, \pi'$
$u \leftarrow g^r$	$\mathsf{pk}_1 \leftarrow g^{\mathsf{sk}_1}, \mathsf{pk}_2 \leftarrow \mathsf{pk}/\mathsf{pk}_1$	$\mathsf{ChE}^{H_1,\tilde{H}_1}[\pi\,|\,{}^{u}_{g}]$
$\pi \leftarrow_s \mathsf{KnE}^{H_1,\tilde{H}_1}[r\,|\,{}^{u}_{g}]$	$\pi' \leftarrow_s \mathsf{KnE}^{H_2,\tilde{H}_2}[\mathsf{sk}_1\,|\,{}^{\mathsf{pk}_1}_{g}|u]$	$\mathsf{ChE}^{H_2,\tilde{H}_2}[\pi'\,|\,{}^{\mathsf{pk}_1}_{g}|u]$
return $\mathsf{pk}^r, (u,\pi)$	$\longrightarrow P_2 : u, \pi, \pi'$	$w \leftarrow u^{\mathsf{sk}_2}$
	$P_2 \longrightarrow: w, \pi''$	$\pi'' \leftarrow_s \mathsf{DHP}^{H_3}[\mathsf{sk}_2\,|\,{}^{u,w}_{g,\mathsf{pk}_2}|\pi']$
	$\mathsf{ChP}^{H_3}[\pi''\,|\,{}^{u,w}_{g,\mathsf{pk}_2}|\pi']$	$\longrightarrow P_1 : w, \pi''$
	return $u^{\mathsf{sk}_1} \cdot w$	return $\top$

Fig. 3. Encryption and decryption for the scheme NPS

We see that the ciphertext is simply an ElGamal ciphertext, together with a simulatable proof of knowledge of the exponent r. When decrypting, the client first verifies this proof. If an assertion fails, then $\perp$ is immediately returned. As next, the client asks the server to apply its private key share sk_2 to the ciphertext u. This request includes a Schnorr proof of knowing the private key sk_1, where the challenge depends on the ciphertext; hence this proof cannot be reused. The request also contains the value h'' that allows the simulator to perform an exponentiation with sk_1. The server verifies the Schnorr proofs of knowing both r and sk_1, and then computes $w \leftarrow u^{\mathsf{sk}_2}$. The value w is returned together with a Schnorr proof that it has been correctly computed—that $\log_u w = \log_g \mathsf{pk}_2$. The client verifies this proof, applies its private key share sk_1 to u, and combines the result with the plaintext share w obtained from the server. It is clear that the scheme satisfies Definition 1 and Definition 4 due to the NIZK proofs.

Theorem 1. *In ROM, if the DDH problem is hard in group* $\mathbb{G}$*, then* NPS *provides IND-CCA against the server and the client.*

Proof. To show IND-CCA against the server, let $\mathcal{A}$ be an adversary that has non-negligible advantage in experiment IND-CCA-S$^{\mathcal{A}}$ with the scheme NPS. We construct an algorithm $\mathcal{S}$ that solves the DDH problem in $\mathbb{G}$. The algorithm $\mathcal{S}$ (called "simulator") internally calls $\mathcal{A}$, realizing the oracles it accesses, including the random oracle. It receives $(h_1, h_2, h_3) \in \mathbb{G}^3$ as an input, and outputs whether they are a DH tuple. In the experiment, the values h_i play the following roles: $\mathsf{pk} = h_1$, $u = h_2$, $k_b = h_3$. We see that if (g, h_1, h_2, h_3) are [resp. are not] a DH tuple, then (pk, u, k_b) are distributed identically to the case $b = 0$ [resp. $b = 1$].

The internal state of $\mathcal{S}$ contains the tables $\mathcal{T}_i, \tilde{\mathcal{T}}_j$ storing the current states of random oracles H_i, $\tilde{H}_j$. For tables $\mathcal{T}_i$, a table row contains the argument made to the oracle, and the given response. For $\tilde{\mathcal{T}}_j$, a table row additionally contains the exponent generated while responding a $\tilde{H}_j$-query.

We now describe how $\mathcal{S}$ behaves in different interactions with $\mathcal{A}$. For key generation, receive pk_2 (committed and opened) from $\mathcal{A}$, while sending it a commitment later opened to $\mathsf{pk}_1 \leftarrow h_1/\mathsf{pk}_2$. For responding a random oracle query (either directly from $\mathcal{A}$, or from simulating the responses to other queries)

$H_i(x)$, look up the row (x, v) from T_i, generating $v \leftarrow_{\$} \mathbb{Z}_p$ and adding the row to the table, if it is not there. Then respond with v. Do the same for $\tilde{H}_2$-query (put $\perp$ as the exponent). Also do the same for query $\tilde{H}_j(x)$, but if the row (x, v, t) is not yet in $\tilde{T}_j$ then generate $t \leftarrow_{\$} \mathbb{Z}_p$, define $v = \mathsf{pk}^{1/t}$, add (x, v, t) to $\tilde{T}_1$, and return v.

Simulator S has to prepare a challenge ciphertext $c = (u, h', \alpha, \alpha', \gamma)$ for A. We have defined u; the rest is constructed by faking the NIZK proofs: S generates $s \leftarrow_{\$} \mathbb{Z}_p$, defines $g' \leftarrow g^s$, $h' \leftarrow h_2^s$, generates $\beta, \gamma \leftarrow_{\$} \mathbb{Z}_p$, computes $\alpha \leftarrow u^\beta / g^\gamma$ and $\alpha' \leftarrow (h')^\beta / (g')^\gamma$, and adds $((g', u, h', \alpha, \alpha'), \beta)$ to T_1 and $((u, \alpha), g', \perp)$ to $\tilde{T}_1$. This computation may fail if the added rows are already present in the tables, but this happens with only negligible probability because all arguments contain fresh randomness.

A query from A to the Dec-oracle with argument $c^* = (u^*, h'^*, \alpha^*, \alpha'^*, \gamma^*)$ is handled by S as follows. First check the proofs, similarly to $\mathsf{NPS}.DC_C$. Return $\perp$, if they fail. Otherwise look up the row $((u^*, \alpha^*), g', t)$ in $\tilde{T}_1$ and return $(h'^*)^t$. This row has to exist, because the proof-checks look it up. The value t may be missing, but in this case c^* had to be the challenge ciphertext.

A query from A to the DC_C-oracle with the same argument c^* is handled by S as follows. Check the proofs; return $\perp$ if they fail. Prepare the query to the server, faking the proof of knowledge π' of sk_1. Whenever A invokes DC_C for the second round, S ignores this query: the server is not expected to get any answer from the client's second round.

Throughout this construction, all values that A sees are distributed identically to the experiment IND-CCA-S^A for NPS. In particular, the responses from the random oracles are uniform and mutually independent. Finally, A gives its guess b^*, which S outputs. The advantage of S is equal to the advantage of A.

The proof of IND-CCA against client is similar. When the A generates the simulatable proof of knowledge of sk_1 (as in DC_C) and invokes $\tilde{H}_2(g, \mathsf{pk}_1, u)$ for that purpose, S chooses u as the value it wants to raise to power sk_1. If A then invokes the oracle DC_S, the simulator S can reply with $w \leftarrow \mathsf{pk}^r / u^{\mathsf{sk}_1}$. $\square$

Theorem 2. *In ROM, if the DDH problem is hard in group $\mathbb{G}$, then NPS provides CCA against offline guessing attacks by the client.*

Proof. Similarly to the proof of Theorem 1, we assume the existence of an adversary A that has advantage at least $1/2 + T/L + \nu$ for a non-negligible ν in the experiment OG-CCA-C$_{T,L}^A$ with the scheme NPS, and construct the algorithm S that solves the DDH problem in $\mathbb{G}$. It again gets (h_1, h_2, h_3) as the input, and again uses them as $\mathsf{pk} = h_1$, $u = h_2$, $k_b = h_3$.

The internal state of S consists of the same tables as in the proof of Theorem 1. For some $\mathsf{h} \in \mathbb{G}$, we additionally define $T_2|_\mathsf{h}$ as the subset of rows of T_2 of the form $((\mathsf{h}, \alpha_1, c), v)$ (i.e. the first component to the argument of H_2 was h). The simulator S also maintains a set of integers $\mathbf{K}$, initialized to $\{1, \ldots, L\}$.

Simulator S has to prepare the arguments to A. The challenge ciphertext $c = (u, h', \alpha, \alpha', \gamma)$ is prepared identically to the proof of Theorem 1. The list of potential private key shares of the client is defined by generating

$\mathsf{sk}_1^{(1)}, \ldots, \mathsf{sk}_1^{(L)} \leftarrow_\$ \mathbb{Z}_p$ and putting $SK_1 = [\mathsf{sk}_1^{(1)}, \ldots, \mathsf{sk}_1^{(L)}]$. Note that the inputs to $\mathcal{A}$ are distributed identically to the experiment OG-CCA-C$_{T,L}^{\mathcal{A}}$ for NPS. Also note that at this point, $\mathcal{S}$ has not selected the "right" private key share. Define also $\mathsf{pk}_1^{(i)} \leftarrow g^{\mathsf{sk}_1^{(i)}}$ for $i \in \{1, \ldots, L\}$.

A query from $\mathcal{A}$ to either one of the hash functions or to the $\mathcal{D}ec$-oracle is handled identically to the proof of Theorem 1. Again, all responses to these queries are distributed identically to the actual experiment OG-CCA-C$_{T,L}^{\mathcal{A}}$.

The t-th query $(c_\mathsf{t}, \alpha_{1,\mathsf{t}}, \gamma_{1,\mathsf{t}})$ to the $\mathcal{DC}_\mathsf{S}$-oracle is handled by $\mathcal{S}$ as follows. First, verify the proofs in $c_\mathsf{t} = (u_\mathsf{t}, h'_\mathsf{t}, \alpha_\mathsf{t}, \alpha'_\mathsf{t}, \gamma_\mathsf{t})$, corresponding to the assertions in NPS.$\mathcal{DC}_\mathsf{S}$. Return $\bot$, if these verifications fail.

The simulator continues with the response for $\mathcal{DC}_\mathsf{S}$-oracle as follows. It finds the index $i \in \{1, \ldots, L\}$, such that the row $((\mathsf{pk}_1^{(i)}, \alpha_{1,\mathsf{t}}, \gamma_{1,\mathsf{t}}), \beta_{1,\mathsf{t}})$ is in the table $\mathcal{T}_2$ for some $\beta_{1,\mathsf{t}}$, and $g^{\gamma_{1,\mathsf{t}}} = \alpha_{1,\mathsf{t}} \cdot (\mathsf{pk}_1^{(i)})^{\beta_{1,\mathsf{t}}}$. If there is no such row, then let $i = \bot$. The probability of having more than one such row in $\mathcal{T}_2$ is negligible. Indeed, if i and i' would both be such indices, then $(\mathsf{pk}_1^{(i)})^{\beta_{1,\mathsf{t}}} = (\mathsf{pk}_1^{(i')})^{\beta'_{1,\mathsf{t}}}$. The values $\beta_{1,\mathsf{t}}$ and $\beta'_{1,\mathsf{t}}$ are random, and generated independently from $\mathsf{pk}_1^{(i)}$ and $\mathsf{pk}_1^{(i')}$, hence this equality can hold only with negligible probability.

If $i \notin \mathbf{K}$, then $\mathcal{S}$ returns $\bot$ to $\mathcal{A}$. If $i \in \mathbf{K}$, then $\mathcal{S}$ has to decide whether the "right" private key is $\mathsf{sk}_1^{(i)}$. For this purpose, $\mathcal{S}$ tosses a biased coin, with the result "heads" having the probability $1/|\mathbf{K}|$. If the result is "heads", then this means that the "right" private key was indeed $\mathsf{sk}_1^{(i)}$. In this case, $\mathcal{S}$ gives up the simulation, outputting $\bot$. Otherwise, $\mathcal{S}$ sets $\mathbf{K} \leftarrow \mathbf{K} \setminus \{i\}$ and returns $\bot$ to $\mathcal{A}$.

Again, we have that as long as $\mathcal{A}$ has not managed to find the "right" private key, all values in the simulation are distributed identically to the experiment OG-CCA-C$_{T,L}^{\mathcal{A}}$ for NPS. The probability of finding the "right" private key is upper-bounded by T/L, hence $\mathcal{S}$ still has at least the non-negligible advantage ν in solving the DDH problem in $\mathbb{G}$. $\square$

5 Fit for Our Main Use-Case

In our main use-case, the client is a smartphone, receiving and storing encrypted messages (e.g. credentials), and decrypting them for short uses. The smartphone communicates with the helper server over mobile internet, through a secure channel. An attacker in this system may have the following goals: (A) learn the plaintext corresponding to a ciphertext, or (B) make the phone accept wrong plaintext. Against this attacker we deploy NPS, including the clone detection. The latter may be continuous.

We consider the following attacks that an attacker may perform: (1) convince the phone to start decryption protocol with a particular ciphertext; (2) learn phone's encrypted memory; (3) learn phone's unencrypted memory; (4) learn server's keyshare; (5) take passive control over server; (6) take active control over server; (7) masquerade as phone to the server; (8) masquerade as server to the phone.

Let the boolean variable x indicate whether the clone detection is done continuously. Let y_i ($i \in \{1, \ldots, 8\}$) be a boolean variable indicating whether adversary has successfully performed the attack (i); note that $y_3 \Rightarrow y_2$, $y_6 \Rightarrow y_5$, and $y_5 \Rightarrow y_4$. Let the boolean variables z_X for $X \in \{A, B\}$ indicate that the adversary achieves his goal X. Let also $\tilde{z}_X$ denote that the goal is achieved only for a short time (until the continuously running clone detection mechanism discovers something); obviously $z_X \Rightarrow \tilde{z}_X$. The following implications hold:

$$[y_2 \wedge y_4] \vee [((y_2 \wedge y_8) \vee y_3) \wedge y_7 \wedge \neg x] \Rightarrow z_A \qquad ((y_2 \wedge y_8) \vee y_3) \wedge y_7 \Rightarrow \tilde{z}_A$$

Also, if some z_X or $\tilde{z}_X$ is true, then this must follow from some of the implications. We see that z_B never holds, because Definition 4 forbids it. Server's keyshare and phone's encrypted keyshare are sufficient for decryption. So is the knowledge of phone's unencrypted keyshare, if the attacker can masquerade the phone and the clone detection does not stop the attack. Interestingly, $y_2 \wedge y_8 \Rightarrow y_3$, because the information sent in phone's first message may enable the offline guessing attack against the PIN. We consider all these vulnerabilities acceptable.

We see that our scheme adds significant overhead to "plain" IND-CPA secure ElGamal. We also see that the overheads are wholly acceptable for our main intended use-case. We have implemented NPS encryption and decryption in Python on top of the PyCryptodome cryptographic library, using the elliptic curve group P-256 as $\mathbb{G}$; the running times are 6 ms for $\mathcal{E}nc$, 135 ms for $\mathcal{DC}_C$, and 107 ms for $\mathcal{DC}_S$ running on a laptop with an Intel® Core™ i5-10210U CPU and 16GB RAM. The size of a key encapsulation is 6 kilobytes, the messages sent from the client to the server and back: 6.7KB and 5.5KB, respectively.

Acknowledgement. This research has been funded by the European Regional Development Fund through EXCITE, the Estonian Centre of Excellence in ICT.

References

1. Asokan, N., Tsudik, G., Waidner, M.: Server-supported signatures. J. Comput. Secur. **5**(1), 91–108 (1997). https://doi.org/10.3233/JCS-1997-5105
2. Baudet, M.: Deciding security of protocols against off-line guessing attacks. In: Proceedings of the 12th ACM Conference on Computer and Communications Security, CCS 2005, pp. 16–25. Association for Computing Machinery, New York (2005). https://doi.org/10.1145/1102120.1102125
3. Bernhard, D., Fischlin, M., Warinschi, B.: On the hardness of proving CCA-security of signed ElGamal. In: Cheng, C.-M., Chung, K.-M., Persiano, G., Yang, B.-Y. (eds.) PKC 2016. LNCS, vol. 9614, pp. 47–69. Springer, Heidelberg (2016). https://doi.org/10.1007/978-3-662-49384-7_3
4. Bicakci, K., Baykal, N.: Server assisted signatures revisited. In: Okamoto, T. (ed.) CT-RSA 2004. LNCS, vol. 2964, pp. 143–156. Springer, Heidelberg (2004). https://doi.org/10.1007/978-3-540-24660-2_12
5. Boneh, D., Shoup, V.: A Graduate Course in Applied Cryptography (2020). A book in preparation, v0.5

6. Brañdao, L.T.A.N., Mouha, N., Vassilev, A.: Threshold schemes for cryptographic primitives. Technical report. NISTIR 8214, National Institute of Standards and Technology (NIST) (2019)

7. Buldas, A., Kalu, A., Laud, P., Oruaas, M.: Server-supported RSA signatures for mobile devices. In: Foley, S.N., Gollmann, D., Snekkenes, E. (eds.) ESORICS 2017. LNCS, vol. 10492, pp. 315–333. Springer, Cham (2017). https://doi.org/10.1007/978-3-319-66402-6_19

8. Buldas, A., Laanoja, R., Truu, A.: A server-assisted hash-based signature scheme. In: Lipmaa, H., Mitrokotsa, A., Matulevičius, R. (eds.) NordSec 2017. LNCS, vol. 10674, pp. 3–17. Springer, Cham (2017). https://doi.org/10.1007/978-3-319-70290-2_1

9. Chaum, D., Pedersen, T.P.: Wallet databases with observers. In: Brickell, E.F. (ed.) CRYPTO 1992. LNCS, vol. 740, pp. 89–105. Springer, Heidelberg (1993). https://doi.org/10.1007/3-540-48071-4_7

10. Cramer, R., Shoup, V.: A practical public key cryptosystem provably secure against adaptive chosen ciphertext attack. In: Krawczyk, H. (ed.) CRYPTO 1998. LNCS, vol. 1462, pp. 13–25. Springer, Heidelberg (1998). https://doi.org/10.1007/BFb0055717

11. De Santis, A., Desmedt, Y., Frankel, Y., Yung, M.: How to share a function securely. In: Proceedings of the Twenty-Sixth Annual ACM Symposium on Theory of Computing, STOC 1994, pp. 522–533. Association for Computing Machinery, New York (1994). https://doi.org/10.1145/195058.195405

12. Ding, Y., Horster, P.: Undetectable on-line password guessing attacks. ACM SIGOPS Oper. Syst. Rev. **29**(4), 77–86 (1995). https://doi.org/10.1145/219282.219298

13. Dodis, Y., Katz, J., Xu, S., Yung, M.: Key-insulated public key cryptosystems. In: Knudsen, L.R. (ed.) EUROCRYPT 2002. LNCS, vol. 2332, pp. 65–82. Springer, Heidelberg (2002). https://doi.org/10.1007/3-540-46035-7_5

14. Proposal for a Regulation of the European Parliament and of the Council amending Regulation (EU) No 910/2014 as regards establishing a framework for a European Digital Identity (SEC(2021) 228 final) - (SWD(2021) 124 final) - (SWD(2021) 125 final) (2021). https://digital-strategy.ec.europa.eu/en/library/trusted-and-secure-european-e-id-regulation

15. ElGamal, T.: A public key cryptosystem and a signature scheme based on discrete logarithms. In: Blakley, G.R., Chaum, D. (eds.) CRYPTO 1984. LNCS, vol. 196, pp. 10–18. Springer, Heidelberg (1985). https://doi.org/10.1007/3-540-39568-7_2

16. Fiat, A., Shamir, A.: How to prove yourself: practical solutions to identification and signature problems. In: Odlyzko, A.M. (ed.) CRYPTO 1986. LNCS, vol. 263, pp. 186–194. Springer, Heidelberg (1987). https://doi.org/10.1007/3-540-47721-7_12

17. Fischlin, M.: Trapdoor commitment schemes and their applications. Ph.D. thesis, Goethe University Frankfurt, Frankfurt am Main, Germany (2001). https://zaurak.tm.informatik.uni-frankfurt.de/diss/data/src/00000229/00000229.pdf.gz

18. ISO 18033-2: Encryption algorithms—Part 2: Asymmetric ciphers. Standard, International Organization for Standardization (2006)

19. Sarr, A.P.: Cryptanalysis and improvement of smart-ID's clone detection mechanism. Cryptology ePrint Archive, Paper 2019/1412 (2019). https://eprint.iacr.org/2019/1412

20. Schnorr, C.P.: Efficient identification and signatures for smart cards. In: Brassard, G. (ed.) CRYPTO 1989. LNCS, vol. 435, pp. 239–252. Springer, New York (1990). https://doi.org/10.1007/0-387-34805-0_22

21. Shoup, V., Gennaro, R.: Securing threshold cryptosystems against chosen cipher-text attack. J. Cryptol. **15**(2), 75–96 (2002). https://doi.org/10.1007/s00145-001-0020-9

22. Vauclair, M.: Secure element. In: van Tilborg, H.C.A., Jajodia, S. (eds.) Encyclopedia of Cryptography and Security, pp. 1115–1116. Springer, Boston (2011). https://doi.org/10.1007/978-1-4419-5906-5_303

Deep Learning for Security and Trust

Software Vulnerability Detection via Multimodal Deep Learning

Xin Zhou[✉] and Rakesh M. Verma[✉]

University of Houston, Houston, TX, USA
xzhou21@uh.edu, rmverma2@central.uh.edu

Abstract. Vulnerabilities in software are like ticking time bombs, but it is difficult to completely eliminate them. For example, buffer overflow is a quite common vulnerability that occurs when a program receives too much data that can corrupt nearby space in memory and manipulate other data for malicious actions. To detect potential vulnerabilities in source code, we consider the code as multisource data by extracting semantically meaningful sub-graphs: Abstract Syntax Tree Graph (ASTG) and Tokenized Data Flow Graph (TDFG). We combine these with the original sequence of tokens and 49 heuristic features to train and leverage a multimodal deep learning network to detect vulnerable statements. We propose a Multisource Deep Learner (MDL) with joint representations based on the pretrained attention-based Bidirectional Gated Recurrent Unit (BGRU) neural networks for vulnerability detection in source code. Our framework not only detects potential vulnerabilities but also locates and ranks the vulnerable statements according to their importance based on the Program Dependence Graph (PDG). Our results show that an MDL-based model using multiple modalities is significantly better than a single modality based model. We also present comparisons with state-of-the-art methods.

Keywords: Static Analysis · Source Code · Software Bugs · Data Flow Graph · Abstract Syntax Tree · Deep Learning

1 Introduction

During the software development and deployment process, the later the bug is found, the greater the cost of repair. Most of the software defects are introduced in the coding stage, some of them escape detection in the current approaches of unit testing, integration testing, functional testing, and acceptance testing. InfoQ [39] reported that 30% to 70% of code logic design and coding defects can be discovered and repaired through static code analysis. Hicken et al. [14] also reported that, as expected, 85% of defects come in during the coding phase, but only a few defects are found during coding since we typically find bugs when we start testing the programs. Static code analysis plays a very critical role in the secure development process, and it must be moved forward as much as

G. Lenzini and W. Meng (Eds.): STM 2022, LNCS 13867, pp. 85–103, 2023.
https://doi.org/10.1007/978-3-031-29504-1_5

possible, since earlier detection can reduce the cost of development and repair for developers and companies. Many companies will likely encounter substantial resistance from developers to implement static code analysis tools due to the large number of false alarms that are generated. This means developers will waste considerable time in bug confirmation. Therefore, only a suitable and practicable static analysis tool can really reduce the development cost. There are two main static code analysis methods: 1) analyze intermediate files compiled from source code such as binary, language-independent intermediate representation (LLVM), etc., and 2) analyze source code directly through semantic information extracted from source files. Our framework is focused on source code itself.

According to [17,19,36], the main techniques for static code analysis are: 1) developing a defect pattern database and then matching the code to be analyzed with common defect patterns to detect potentially vulnerable statements. This method is simple and convenient but needs enough patterns and is typically prone to false positives. 2) Type inference refers to the automatic detection of the type of an expression in a formal language to ensure that each statement in the code has the correct type. 3) Model checking is based on finite state automata. The impact of each statement is abstracted into a state of a finite state automaton, and then the purpose of code analysis is achieved by analyzing the finite state machine. It can check timing characteristics such as program concurrency. 4) Data flow analysis by collecting semantic information from source code and abstracting it with a control flow graph. It can analyze and discover the behaviors of the program during run-time without actually running the program. 5) Data driven prediction using machine learning by utilizing the above analytical techniques based on a large training set that contains a diverse set of vulnerable and non-vulnerable patterns. We focus on data driven techniques.

Multi-modal learning involves relating information from multiple sources such as images and text. Multi-modal representation learning tries to eliminate redundancy and utilizes complementarity between modalities, so as to learn better features representation. Currently, there are two research directions in multi-modal learning: 1) joint representation, which refers to mapping the information of multiple modalities together into a unified multimodal vector space; 2) coordinated representations, which refers to mapping each modality to its respective representation space, but certain correlation constraints (such as linear correlation) are satisfied between the mapped vectors.

In computer vision, multi-modal learning has grown rapidly recently. Unstructured data can inherently take many forms such as visual and textual content. In this paper, we construct two type of modalities, i.e., sequential and graphical representations, from raw data using different constructors. Then, we focus on static vulnerability detection in source code via multi-modal learning and make the following contributions:

1. We propose a new tokenization method with abstract representation of numbers that outperforms state of art methods in rigorously repeated experiments with random train, valid, and test dataset splits and averaged results.
2. We create a multi-modal dataset for vulnerability detection in source code.

3. We propose the Multisource Deep Learner for vulnerability detection in source code via multi-modal learning.
4. We propose the Vulnerability Highlighter to locate vulnerable statements and rank the relevant statements.
5. We conduct a series of ablation experiments to show the value of significant components of our ML pipeline.

Organization. After the Related Work section, in Sect. 3, we explain how we extract and tokenize source code as four modalities from different perspectives. Section 4 details the data-driven prediction method that learns code patterns and dependency graph to detect the vulnerabilities and locate vulnerable statements. In Sect. 5, we describe the datasets used for the evaluation. Experimental details and results are discussed in Sect. 6 and Sect. 7 concludes the paper.

2   Related Work

We discuss the related work on this topic in four categories: custom token-based approaches, abstract syntax tree based approaches, data driven approaches, and multimodal learning based approaches.

Custom Token-Based Approaches: Russell et al. [33] design a function-level vulnerability detection system using machine learning. They compile millions of open-source functions and label them with carefully selected findings from three different static analyzers that indicate potential exploits. The authors have applied a variety of ML techniques inspired by classification problems in the natural language domain, fine-tuned them for their application, and achieved the best overall results via convolutional neural network and classified with an ensemble tree algorithm. However, function-level vulnerability detection is not as useful as statement-level detection in real-world detection, since functions can be too large (e.g., 4,000 and 12,000 line functions are mentioned in [25]) and time-consuming for an expert to manually investigate.

Abstract Syntax Tree-Based Approaches: Mark Weiser [40] designed a program slicing method for automatically decomposing programs by analyzing their data flow and control flow. The author mentions this program slicing method can be used for debugging and parallel processing of slices. Recently, several automatic vulnerability detection works are based on a similar idea of combining data flow, control flow, and Abstract Syntax Tree (AST).

VulDeePecker [24] is the first system showing the feasibility of using deep learning to detect vulnerabilities while being able to narrow down locations of vulnerabilities. The authors also present the first vulnerability dataset for deep learning approaches. VulDeePecker is only able to deal with vulnerabilities related to library/API function calls. Their newer framework SySeVR [23] is used to detect vulnerabilities in source code based on so-called Syntax-Semantics Vector Representation, which is extracted with known potential vulnerable characteristics related to function calls, array usage, pointer usage, or arithmetic expressions. They truncate or pad input as a set of fixed length sequences of tokens (threshold = 500) for neural networks.

The Vulnerability Deep learning-based Locator, VulDeeLocator [22], uses a deep learning-based fine-grained vulnerability detector for C source code. The authors detected four vulnerabilities that were not reported in the National Vulnerability Database (NVD), but their framework is limited to C programs and heavily relies on the LLVM compiler, since their representations are based on the LLVM intermediate representation.

Alon et al. designed a neural model, Code2Vec [2], for representating snippets of code as continuous distributed vectors. They demonstrate the effectiveness of their model to predict a method's name from the vector representation of its body based on the AST. However, their model is only able to predict labels that were observed as-is at training time and unable to compose such names and usually catches only the main idea. This paper inspired us to extend program text representation with different kinds of graph representations.

Other Data Driven Approaches with Different Features: Harer et al. [11] design a software vulnerability detection framework, which is a data-driven approach to detect vulnerabilities with machine learning in C and C++ programs. They use features based on the operations in each basic block (opcode, vector, or op-vecv) derived from a program build process using Clang and LLVM. Then, they combine this with source-based features using C/C++ lexer to predict vulnerability at the function level. Their work is limited by the labels of functions, since it is really hard to manually investigate and validate labels that are generated by other static analysis tools such as Clang static analyzer. Li et al. [21] present a vulnerability detector, based on sub-graphs in the Program dependence Graphs, that outputs the crucial statements that are relevant to the detected vulnerability.

Multimodal Learning Approaches: Heidbrink et al. [6,12,13] proposed a method that uses multimodal learning for flaw detection in software programs based on two modalities (source code and program binary). In source code, they extract subgraph information by counting all unique node-edge-node transitions and flaw analysis-inspired statistical features associated with following program constructs: function call (e.g., number of external calls), variables (e.g., number of explicitly defined variables), graph node counts (e.g., number of else statements), graph structure (degrees of AST nodes by type). For binaries, they used Ghidra to extract and collect statistical count information per function associated with function call, variables (e.g., number of stack variables), function size (e.g., number of basic blocks), and p-code opcode instances, which is Ghidra's intermediate representation language for assembly language instructions.

Other Approaches: In computer vision research improved model have been proposed based onmulti-view techniques. This line of research shows that analyzing an object from different perspectives can extract more semantic features and information. Jin et al. [15] proposed a method to take joint-embedding of shapes and contours. Lai et al. [20] introduced a large-scale, hierarchical multi-view object dataset RGB-D (Red-Green-Blue-Depth) collected using an RGB-D camera. RGB-D based object combines color and depth information to substantially improve results. Mokhov [26] designs a machine learning approach for static code analysis and fingerprinting for security bugs using the MARFCAT [27] application [10]. Sestili et al. [35] points towards future approaches that may solve vulnerability detection problems using representations of code that can capture appropriate scope information and using deep learning methods that are able to perform arithmetic operations. They developed a code generator to produce an arbitrarily large number of code samples of controlled complexity. They also investigated the limits of the current state-of-the-art AI system for detecting buffer overflows and compared it with current static analysis engines. Their data are simple C-like programs, which are generated as basic blocks without loops, conditionals, and variables with unknown value. Katz et al. [16] design a framework to convert a program in low-level representation back to a higher-level human-readable representation based on neural machine translation. Their framework can automatically learn a decompiler from a given compiler. However, their framework fails if the input is longer than the threshold value. Wang et al. [38] propose a graph neural network assisted data flow analysis method to find potential buffer overflows in execution traces. Yamaguchi et al. [41] employ the concept of code property graph in many graph databases such as ArangoDB, Neo4J, and OrientDB and demonstrate its efficacy by identifying 18 previously unknown vulnerabilities in the source code of the Linux Kernel.

3 Background and Approach

In this section, we first describe and explain how to extract and tokenize source code into different representations as different modalities. Second, we introduce and explain our framework for vulnerability detection in source code.

3.1 Data Representations

These four data representations are the modalities for multimodal learning.

1. Token: we extract and tokenize the sliced code into a sequence of lexical tokens based on the Program Dependence Graph (the definition is in Sect. 3.4).
2. Abstract Syntax Tree Graph (ASTG): is a graph type modality, which is generated by AST constructor.
3. Tokenized Data Flow Graph (TDFG): is a graph type modality, which is based on data flow dependencies.
4. Heuristic Features: the syntactic complexity properties of source code [4] (e.g., number of variable operations, number of function calls, etc.). Totally, we have 49 features [4] generated from the properties of AST and tokens.

ASTG and TDFG are extracted as structural semantic information similar to depth scans for images in computer vision. For example, when you consider a specific variable in source code, you focus only on the lines that use this variable.

3.2   Potential Vulnerable Statement

Potential Vulnerable Statement is a pre-defined collection from Li et al. [23] based on the Checkmarx over open-source tools Flawfinder [9] and RAT [31]. This collection is used for extracting program dependence graph and highlighting the vulnerabilities.

3.3   Abstract Syntax Tree

An AST is used to represent the abstract syntactic structure of source code in a formal language. Once we have the tree representation of source code, we can mine all possible paths through terminal-to-terminal, root-to-terminal, or other efficient kernels. We use an open-source tool ASTminer [18] to generate the ASTs. Then, we keep the same node ID for the same variables by merging all of them into one node to connect all edges for final AST Graph.

3.4   Program Dependence Graph

Program dependence graph (PDG) [8] consists of control dependency and data dependency, which are defined based on the Control Flow Graph (CFG).

Control Flow Graph (CFG)) [8]: For static analysis, the CFG is essential to extract semantic features and accurately represent the flow inside of a program unit. Let P be a program that consists of functions. The CFG of function f_i is a graph $G_i = (V_i, E_i)$, where V_i is a set of nodes, each node represents a statement or control predicate, and E_i is a set of directed edges such that each edge represents the possible flow of control between a pair of nodes.

Data Dependency [8]: Let P be a program that consists of functions and let the CFG for function f_i be $G_i = (V_i, E_i)$. A node n_{ik} will be considered as data dependent if there is a path from n_{ik} to n_{ij} in G_i and a value computed at node n_{ik} is used at node n_{ij}, where $1 \leqslant j, k \leqslant l_i$ and $j \neq k$, where l_i is total number of statements from f_i.

Control Dependency [8]: Let P be a program consisting of functions f_i with CFG $G_i = (V_i, E_i)$. If there exists a path starting at n_{ik} and ending at n_{ij} such that (i) n_{ij} post-dominates every node on the path excluding n_{ik} and n_{ij}, and (ii) does not post-dominate n_{ij}, then n_{ij} is control dependent on n_{ik}.

PDG [8]: Let P be a program that consisting of functions f_i with PDG $G_i' = (V_i, E_i')$, where V_i is the same as G_i for CFG and E' is a set of directed edges such that each edge represents a data or control dependency between a pair of nodes.

3.5   TDFG and ASTG

Tokenized Data Flow Graph (TDFG) is constructed based on the tokenized program by the following steps: 1) collect potential vulnerable statement line numbers, 2) generate data flow graph based on these collected line numbers, 3) construct a graph G = (V, E) with tokenized source code where a node v_i represents a partial statement and an edge represents a the flow of data between a pair of nodes. Final TDFG feature set is a collection of sub-graphs from TDFG based on the potential vulnerable statements. Our previous work [42] shows how the TDFG is constructed and how sub-graphs are extracted. Abstract Syntax Tree Graph has the same extraction process as TDFG. Since AST tree can be directly represented as G = (V, E) where V is a set of nodes and E is a set of edges, where a node represents a token type and a edge represents a possible flow of control between a pair of nodes. We convert each potential vulnerable statement as a shared node (using same node index) over all modalities for alignment. Both TDFG and ASTG sequences and sub-graphs can be embedded as word or graph level embedding.

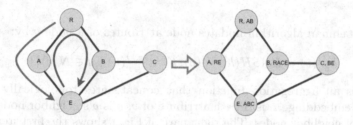

Fig. 1. A sub-graph sample. R is root node and E is exit node; Red arrow line is terminal to terminal path and blue arrow line is root to terminal path on the left; the right graph shows the first iteration of WLGK algorithm (Color figure online)

3.6 Sub-graph Extractions

We collect sub-graphs using the following three extraction methods to find semantic representations of source code:

1. Root-to-terminal (RTT): is a collection of paths from the root node to a terminal node.
2. Terminal-to-terminal (TTT): is a collection of paths from a terminal node to a terminal node. This method has been used by Code2Vec [2] and Code2Seq [1].
3. Weisfeiler-Lehman Graph Kernels (WLGK): [37] is a rapid feature extraction scheme based on the Weisfeiler-Lehman test of isomorphism on graphs. We use WLGK to walk through the paths and extract sub-graphs from both ASTG and TDFG since it has been found useful in other tasks, e.g., Graph2Vec [28].

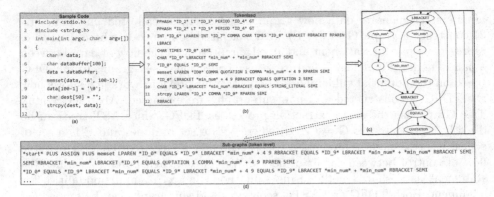

Fig. 2. TDFG sub-graphs extraction example: (a) is a sliced sample for the model, (b) is tokenized PDG, (c) is TDFG, and (d) is a set of sub-graphs in token-level.

Weisfeiler-Lehman algorithm updates node attributes of a node v by:

$$h_i^{(t)}(v) = HASH(h_i^{(t-1)}(v), F\{h_i^{(t-1)}(u) \mid u \in N(v)\})$$ (1)

where F is an aggregation function that concatenates topologically ordered neighbor's embedding, h_i is the ith attribute of v, u is v's neighbor node, and N is the set of neighbor nodes. The right part of Fig. 1 shows the first iteration of Weisfeiler-Lehman algorithm based on the left graph. For our sub-graph extraction, we collect the paths based on 1-dimensional Weisfeiler-Lehman algorithm with 5 iterations (after grid search from 1 to 10).

Representation: we extract and concatenate the sub-graphs as final representation (MAX = 500 tokens) using above methods based on the TDFG and ASTG. Figure 1 is an example of how sub-graphs are extracted by these three methods.

Figure 2(d) is an example of how a sequence of tokens is generated from raw sample code: line 1 is an example of an RTT path, line 2 is an example of TTT, and line 3 is an example of WLGK path in the sub-graphs.

3.7 Pipeline

We propose a multimodal learning framework for vulnerability detection in source code based on different modality extraction methods. Figure 3 shows the overview of our framework. We first generate Abstract Syntax Tree from source code and Program Dependence Graph from tokenized code. Then, we extract sub-graphs from AST as ASTG modality, tokens from PDG as Token modality and heuristic features (HF) from PDG as HF modality and extract sub-graphs from tokenized data flow graph as TDFG modality. The neural network could be any kind of multimodal leaning network to concatenate and align all modalities for final classification.

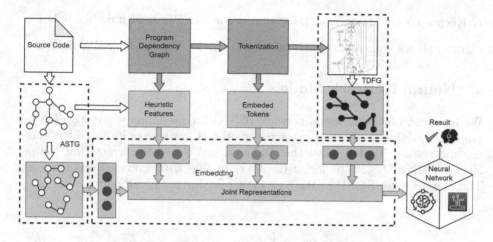

Fig. 3. Multisource Deep Learner Pipeline

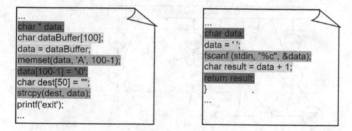

Fig. 4. Vulnerability Highlighter is used to locate vulnerable statements; left example shows stack-based buffer overflow and the right example shows integer overflow.

3.8 Vulnerability Highlighter

We consider the pre-defined potential vulnerable tokens as the Most Possible Vulnerable Statements (MPVSs). If a program is detected as GOOD, we output the result without any highlights. If a program is detected as BAD, we proceed as below:

1. Denote all statements that contain MPVS label as M.
2. Generate control flow graph (CFG), data dependency, and control dependency to construct a program dependence graph (PDG) for each MPV.
3. Label all MPVSs ($[m_1, m_2, ..., m_n] \subseteq M$) with red (dangerous) background color in the program if it is detected as vulnerable.
4. For $i = 1$ to n, we extract their data and control dependencies for CFG G_i.
5. Union all forward slices as one forward list and all backward slices as one backward list respectively for data and control dependencies.
6. Label the statements with orange (warning) background color for backward data dependents of the MPVs if it is not in dangerous. Label other statements from PDG with blue (likely neutral) background color.

7. Keep all other statements, those that are not in PDG, with no highlights.

Figure 4 shows vulnerable statements found by the Vulnerability Highlighter.

4   Neural Network Models

We used convolutional neural network (CNN) for preliminary investigation on graph embedding and feature extraction methods, because of its speed for training and testing. Table 3 shows that Bidirectional Gated Recurrent Unit Neural Network (BGRU) [34] performs the best, in line with previous observations. Therefore, we use BGRU as the base model for further investigation.

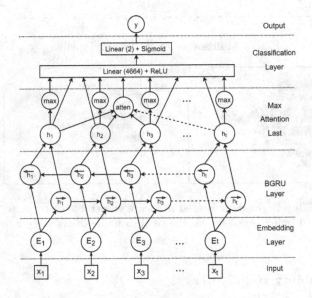

Fig. 5. Attention-based BGRU Classifier

4.1   Attention-Based BGRU

A Bidirectional GRU, or BGRU, is a sequence processing model that consists of two GRUs. One taking the input in a forward direction, and the other in a backwards direction. Gated recurrent units (GRUs) are a gating mechanism in recurrent neural networks, introduced by Kyunghyun Cho et al. [5]. Figure 5 shows how an attention-based BGRU classifier is constructed. Input can be either Token, ASTG, or TDFG. We use pretrained Word2Vec as embedding layer for each modality. A dot product attention layer is followed by BGRU layer. Then, we concatenate the output from attention layer, last hidden layer from BGRU, and max values of all elements from output of the last hidden layer of BGRU as a joint representation for final linear classifier with ReLU and Sigmoid activation functions. Our dot product attention layer is computed as follows:

$$a_t(s) = softmax((\frac{\exp(h_t^{\top} \overline{h}_s)}{\sum_{s'} \exp(h_t^{\top} \overline{h}_s)})$$

where a_t is output representation, s is input vector, and $\overline{h}_s$ is each source hidden state corresponding to the hidden target state h_t (Fig. 5).

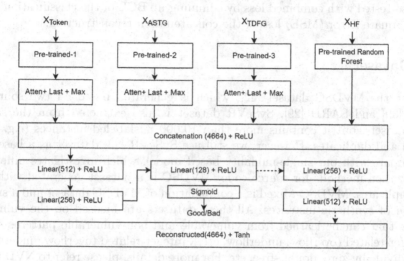

Fig. 6. Multisource Deep Learner

4.2 Multisource Deep Learner

We use three pretrained embedding layers and attention-based BGRU layer as the encoders for token, ASTG, and TDFG modalities. Then, we unfreeze pretrained encoders (learned parameters can still be updated with 0.0001 learning rate) for correlational joint representations (size = 4,664) for vulnerability detection using our Multisource Deep Learner.

Multisource Deep Learner (MDL): it has a similar architecture as Correlational Neural Network (CorrNN [3]), but we use cross entropy loss function instead and added a classifier to fit our classification task. Our framework does not reconstruct all raw inputs, it reconstructs the joint representation by simple MLP encoder-decoder model to get semi-reconstruction loss to fine-tuning classification model using learning rate 0.0001. First, we take concatenated vector [x1, x2, x3, x4] of size d1 + d2 + d3 + d4 from the pooled layer (Attention + Last + Max) based on three pre-trained BGRU and one pre-trained random forest. Given z = (x1, x2, x3, x4), the first hidden layer computes an encoded representation as

$$h_1(z) = f(w_1x_1 + w_2x_2 + w_3x_3 + w_4x_4 + b) \qquad (2)$$

where w is a projection matrix and b is bias vector. Function f can be any non-linear activation function. We grid searched to find the best activation function ReLU for our framework. Our latent vector h is used for classification. We use Binary Cross Entropy (BCE) loss for training. BCE is computed as follows:

$$BCE = -(y \log(p) + (1 - y) \log(1 - p)) \tag{3}$$

where log is the natural log, y is binary indicator and p is predicted probability. We also tested with combined loss by summing up BCE of the classification and Mean Square Error (MSE) loss of the concatenation reconstruction.

5   Dataset

We use the MVDSC dataset [42], which is generated based on two sources: NVD [30] and SARD [29]. SySeVR dataset is also extracted from the same raw datasets, but it contains more than 10,000 mislabeled instances (e.g., see Fig. 7) and duplicates. However, we still use SySeVR [23] dataset as a baseline to compare with our single-modality based model, which investigates different tokenization methods for source code. MVDSC is a dataset generated without any duplicates. MVDSC-Mixed is a combination of MVDSC dataset and a small portion of synthetic instances. All these datasets are focused on the vulnerabilities that can be learned from vulnerable and non-vulnerable patterns such as buffer-related (overflow, underflow, etc.), integer-related (overflow, underflow, etc.), divide-by-zero, double-free, etc. For more details, please refer to NVD [30], SARD [29], and MVDSC [42].

```
static void badSink ( char * data )
char dest [ 50 ] = " " ;
memmove ( dest , data , strlen ( data ) * sizeof ( char ) ) ;
dest [ 50 - 1 ] = ' \0 ' ;
```

```
static void goodG2BSink ( char * data )
char dest [ 50 ] = " " ;
memmove ( dest , data , strlen ( data ) * sizeof ( char ) ) ;
dest [ 50 - 1 ] = ' \0 ' ;
```

Fig. 7. Two code snippets from SySeVR dataset that are identical except for function names, but they label the left as vulnerable, and the right one as non-vulnerable.

5.1   Preprocessing and Tokenization

Each program consists of one or more functions in NVD [30] and SARD [29]. Each function contains labels and comments about vulnerability details including how to fix. Therefore, we need to mask or remove sensitive information that may benefit models. We convert all file names and any token, that contains "bad", 'good', or 'cwe' sub-string (cwe_* contains sensitive information about vulnerabilities), to a fixed common string *C* with star symbols around to avoid code conflicts. We also convert all strings with single quotation mark as '*SQ* + n' and double quotation mark as '*DQ* + n' where n is the length of content in quotation. In addition, we remove all comments. We are using 811 pre-defined

Table 1. Dataset statistics (vulnerable: non-vulnerable).

Dataset	train	valid	test
SySeVR	pool: 64403 (13603:50800)		
MVDSC	7569:22416	1914:5580	1857:5637
MVDSC-Mixed	11416:26569	2401:6093	2325:6169

vulnerable syntax characteristics (memset, strcpy, etc.) which is generated by Li et al. [23] since we use the same raw dataset. We use pycparser [7] as our base lexer to find identifiers including variables and functions (finding identifiers can be tricky). We convert all variable and function names into more semantic meaningful representations (Table 1).

Locate_ID: for masking variable and function names, we need to index them. To keep the index order meaningful, we always index destination (sink) variable before source variable. Ex. strcpy(dest, src) will always be masked as strcpy(*ID_0*, *ID_1*), no matter which variable was declared first. To align those variables which are related to potential vulnerable statement, we denote the variables which are the closest to a potential vulnerable statement starting from 0. That means we can ensure that '*ID_0*' and '*ID_1*' are the two aligned tokens to vulnerable statements since most of function calls take two arguments in our dataset. A more complex function with more arguments can also be handled.

Abstract: after 'Locate_ID', we tokenize the remaining program units based on their types. Once the whole PDG is tokenized, we apply a number abstraction function, **Abstract()**, to convert numbers as (*MIN*, difference) in data flow statements only based on the potential vulnerable statement, where *MIN* represents the minimum number value of all numbers in these data dependents.

6 Experimental Results and Analysis

We now present the results of our experiments and ablation studies.

Metrics. We use accuracy (A), precision (P), recall (R), and F1 as our evaluation metrics. Our dataset is highly skewed since vulnerable statements are far fewer than non-vulnerable statements, so we add extra metric Matthews Correlation Coefficient (MCC) for evaluation.

Comparing Single-Modality Model with Baseline: we use SySeVR [23] as our baseline for single-modality model, since it was developed for detection originally from the same sources as MVDSC dataset [42]. SySeVR dataset contains 64,403 instances and the authors reported their results based on randomly picked dataset 30000/7500/7500 as train/valid/test, we also randomly picked with random seed from the pool with same split ratio. *We report average and*

Table 2. Tokenization comparison using SySeVR dataset; 10 run in 10 different random seeds. SySeVR-BGRU [23] was the best previous result. T is our tokenization method.

Method	A	P	R	F1	MCC
SySeVR-BGRU	94.7	91.5	n/a	86.8	83.6
T + CNN	94.7±0.4	87.6±2.1	87.5±1.7	87.5±0.9	84.2±1.1
T + BGRU	95.3±0.2	90.5±1.7	87.2±2.4	88.8±0.8	85.9±0.9

standard deviation in 10 runs with 10 random seeds, since it is a better evaluation method [32]. Table 2 shows that our single-modality model with same BGRU model as theirs is significantly better than their best result. Hence, we only use the MVDSC dataset [42] for further investigation. For the following experiments, we report the average of three runs in the same train/valid/test sets.

Table 3. Comparing models with token modality on MVDSC dataset [42]

Network	A	P	R	F1	MCC
CNN	95.5	90.4	91.4	90.9	87.9
LSTM	91.9	85.5	81.1	83.3	78.0
BLSTM	95.8	92.1	90.7	91.4	88.6
GRU	96.1	94.9	89.2	91.9	89.5
BGRU	96.6	94.3	91.6	93.0	90.7

Comparing token modality with different models: to find the best model for single-modality and build some pretrained models, we evaluated our token modality with five common networks on the MVDSC dataset. Table 3 shows that the Bidirectional-GRU classifier achieved the best performance among CNN, LSTM, Bidirectional-LSTM, and GRU. The table also shows that both bidirectional LSTM and GRU are better than LSTM and GRU respectively. This suggests that both backward and forward paths are useful for vulnerability detection.

Table 4. Comparing graph embedding in TDFG and ASTG

TDFG2Vec	A	P	R	F1	MCC
token	90.0	82.1	76.6	79.2	72.8
graph	84.7	76.7	55.0	64.1	55.9
ASTG2Vec	Acc	Pre	Recall	F1	MCC
token	91.5	83.7	81.5	82.6	77.0
graph	84.8	72.1	63.0	67.2	57.6

Comparing embedding methods for graph modality: we compare sub-graph embedding and Graph2Vec [28]. For sub-graph embedding, we concatenate all extracted paths as a long sequence (MAX = 500 tokens) and then use a Word2Vec embedding + BGRU (Fig. 5) that is connected with a dot product attention layer for classification. For graph embedding, we use a standard Graph2Vec [28] to embed ASTG or TDFG into a 1024-dimension vector with 5 Weisfeiler-Lehman iterations, then normalize it as a 32×32 grey scale image with a standard CNN classifier. Table 4 shows that token level embedding method is significantly better than graph level embedding. So, we embed a set of sub-graphs as 500 × 32 matrix for further experiments.

Table 5. Comparing tokenization methods on MVDSC dataset

Normal	Locate ID	Abstract	A	P	R	F1	MCC
✓			96.1	94.0	89.8	92.0	89.5
✓	✓		96.3	94.6	89.8	92.4	90.0
✓	✓	✓	96.6	94.3	91.6	93.0	90.7

Comparing tokenization with add-ons: Table 5 shows the differences between different tokenization methods. Two add-ons (Locate_ID and Abstract) eventually and slightly improved the model. With the abstract representation of numbers, the recall is increased by 1.8 which is a critical improvement in vulnerability detection since the size is very sensitive in memory allocation such as malloc()→free().

Table 6. Freezing vs Unfreezing the parameters of pre-trained models

Method	A	P	R	F1	MCC
4 modalities + BCE + freeze	97.0	96.3	91.3	94.8	91.8
4 modalities + BCE + unfreeze	97.7	97.2	93.4	95.2	93.8
4 modalities + CombinedLoss + freeze	95.1	98.5	81.5	89.2	86.7
4 modalities + CombinedLoss + unfreeze	97.8	97.0	93.9	95.4	94.0

Comparing freeze/unfreeze: we compared multiple modalities with frozen and unfrozen mode and tested with two loss functions. Table 6 shows that both unfrozen encoders worked significantly better than their frozen ones. Combined-Loss is not significantly different from BCE but made model training much slower. Therefore, we use BCE for further comparisons. We can see that unfreezing the parameters of the pre-trained model is a better way for fine-tuning.

Table 7. Ablation study of modalities on MVDSC dataset

Modalities	A	P	R	F1	MCC
Token	96.6	94.3	91.6	93.0	90.7
Token + ASTG	97.1	92.3	96.1	94.2	92.2
Token + ASTG + TDFG	97.6	96.9	92.9	94.8	93.2
Token + ASTG + TDFG + HF	97.7	97.2	93.4	95.2	93.8

Comparing single modality and multiple modalities using MVDSC dataset: this ablation study is used to learn how modalities can be stacked up and improve the classification performance in MVDSC dataset. Table 7 shows that all combined model has the best overall performance. The result also shows that ASTG is the booster for higher recall. TDFG and HF make the model more balanced for precision. Comparing token-modality to four combined modalities, the MCC is increased by 3.1% which is significantly better.

Table 8. Model comparisons using MVDSC-Mixed dataset

Modalities	A	P	R	F1	MCC
Token	94.2	94.8	83.6	88.8	85.3
Token + ASTG	95.2	91.0	92.0	91.3	88.0
Token + ASTG + TDFG	95.6	94.1	89.6	91.8	88.9
Token + ASTG + TDFG + HF	95.5	92.7	90.8	91.7	88.7

Table 9. MVDSC vs MVDSC-Mixed

Representations	MVDSC		MVDSC-Mixed		Differences	
	R	MCC	R	MCC	ΔR	ΔMCC
Token	91.6	90.7	83.6	85.3	-8.0	-5.4
Token + ASTG	96.1	92.2	92.0	88.0	-4.1	-4.2
Token + ASTG + TDFG	92.9	93.2	89.6	88.9	-3.3	-4.3
Token + ASTG + TDFG + HF	93.4	93.8	90.8	88.7	-2.6	-5.1

Comparing single modality and multiple modalities using MVDSC-mixed dataset: MVDSC-Mixed adds around 10% adversarial data to MVDSC. Table 8 shows that all modalities are negatively impacted by adversarial data. Table 9 shows that single-modality based model is the most negatively impacted to both recall (−8.0%) and MCC (−5.4%). Therefore, using multiple modalities not only improves the detection performance but also improves the robustness of the model.

7   Conclusion

We propose Multisource Deep Learner, a multimodal learning framework to detect vulnerabilities in source code and show their location in code. The framework mines semantic information for developers. We compared our framework with state-of-the-art algorithms from previous works. We evaluated our system with our multi-modal dataset MVDSC [42]. Our results show that multi-modality-based models are significantly better in performance and robustness than single-modality-based models by the dataset-based evaluation.

Acknowledgments. Research partially supported by NSF grants 1433817 and 2210198, ARO grant W911NF-20-1- 0254, and ONR award N00014-19-S-F009. Verma is the founder of Everest Cyber Security and Analytics, Inc.

A   Appendix

A.1   Limitations

Apart from the usual limitations of static analysis and machine learning, other limitations are: 1) adversarial data may negatively impact model's performance, 2) the current implementation does not address interprocedural analysis.

References

1. Alon, U., Brody, S., Levy, O., Yahav, E.: Code2seq: generating sequences from structured representations of code. In: International Conference on Learning Representations (2019). https://openreview.net/forum?id=H1gKYo09tX
2. Alon, U., Zilberstein, M., Levy, O., Yahav, E.: Code2vec: learning distributed representations of code. Proc. ACM Program. Lang. **3**(POPL) (2019). https://doi.org/10.1145/3290353
3. Chandar, S., Khapra, M.M., Larochelle, H., Ravindran, B.: Correlational neural networks. Neural Comput. **28**(2), 257–285 (2016). https://doi.org/10.1162/NECO_a_00801
4. Chernis, B., Verma, R.: Machine learning methods for software vulnerability detection. In: Proceedings of the Fourth ACM International Workshop on Security and Privacy Analytics, pp. 31–39 (2018)
5. Chung, J., Gulcehre, C., Cho, K., Bengio, Y.: Empirical evaluation of gated recurrent neural networks on sequence modeling. In: NIPS 2014 Workshop on Deep Learning, December 2014 (2014)
6. Cooper, A., Zhou, X., Heidbrink, S., Dunlavy, D.M.: Using neural architecture search for improving software flaw detection in multimodal deep learning models. arXiv:2009.10644 (2020)
7. Eliben: Complete c99 parser in pure python: pycparser v2.21. https://github.com/eliben/pycparser/blob/master/pycparser. Accessed Nov 2021
8. Ferrante, J., Ottenstein, K.J., Warren, J.D.: The program dependence graph and its use in optimization. ACM Trans. Program. Lang. Syst. (TOPLAS) **9**(3), 319–349 (1987). https://doi.org/10.1145/24039.24041

9. Flawfinder: Flawfinder. https://dwheeler.com/flawfinder/. Accessed Feb 2022
10. SQ Group: Static analysis tool exposition (SATE) VI workshop. https://www.nist.gov/itl/ssd/software-quality-group/static-analysis-tool-exposition-sate-vi-workshop. Accessed Mar 2022
11. Harer, J.A., et al.: Automated software vulnerability detection with machine learning. arXiv abs/1803.04497 (2018)
12. Heidbrink, S., Rodhouse, K.N., Dunlavy, D.M.: Multimodal deep learning for flaw detection in software programs. arXiv:2009.04549 (2020)
13. Heidbrink, S., Rodhouse, K.N., Dunlavy, D., Cooper, A., Zhou, X.: Joint analysis of program data representations using machine learning for improved software assurance and development capabilities (2020). https://doi.org/10.2172/1670527. https://www.osti.gov/biblio/1670527
14. Hicken, A.: The shift-left approach to software testing. https://www.stickyminds.com/article/shift-left-approach-software-testing. Accessed Mar 2022
15. Jin, A., Fu, Q., Deng, Z.: Contour-based 3D modeling through joint embedding of shapes and contours. In: Symposium on Interactive 3D Graphics and Games, I3D 2020. Association for Computing Machinery, New York (2020). https://doi.org/10.1145/3384382.3384518
16. Katz, O., Olshaker, Y., Goldberg, Y., Yahav, E.: Towards neural decompilation. arXiv abs/1905.08325 (2019)
17. Kotenko, I., Izrailov, K., Buinevich, M.: Static analysis of information systems for IoT cyber security: a survey of machine learning approaches. Sensors **22**(4) (2022). https://doi.org/10.3390/s22041335. https://www.mdpi.com/1424-8220/22/4/1335
18. Kovalenko, V., Bogomolov, E., Bryksin, T., Bacchelli, A.: PathMiner: a library for mining of path-based representations of code. In: Proceedings of the 16th International Conference on Mining Software Repositories, pp. 13–17. IEEE Press (2019)
19. Kulenovic, M., Donko, D.: A survey of static code analysis methods for security vulnerabilities detection. In: 2014 37th International Convention on Information and Communication Technology, Electronics and Microelectronics (MIPRO), pp. 1381–1386 (2014). https://doi.org/10.1109/MIPRO.2014.6859783
20. Lai, K., Bo, L., Ren, X., Fox, D.: A large-scale hierarchical multi-view RGB-D object dataset. In: 2011 IEEE International Conference on Robotics and Automation, pp. 1817–1824 (2011). https://doi.org/10.1109/ICRA.2011.5980382
21. Li, Y., Wang, S., Nguyen, T.N.: Vulnerability detection with fine-grained interpretations, pp. 292–303. Association for Computing Machinery, New York (2021). https://doi.org/10.1145/3468264.3468597
22. Li, Z., Zou, D., Xu, S., Chen, Z., Zhu, Y., Jin, H.: VulDeeLocator: a deep learning-based fine-grained vulnerability detector. IEEE Trans. Dependable Secure Comput. **19**(4), 2821–2837 (2022). https://doi.org/10.1109/TDSC.2021.3076142
23. Li, Z., Zou, D., Xu, S., Jin, H., Zhu, Y., Chen, Z.: SySeVR: a framework for using deep learning to detect software vulnerabilities. IEEE Trans. Dependable Secure Comput. 1 (2021). https://doi.org/10.1109/tdsc.2021.3051525
24. Li, Z., et al.: VulDeePecker: a deep learning-based system for vulnerability detection. In: 25th Annual Network and Distributed System Security Symposium, NDSS 2018, San Diego, California, USA, 18–21 February 2018. The Internet Society (2018). http://wp.internetsociety.org/ndss/wp-content/uploads/sites/25/2018/02/ndss2018_03A-2_Li_paper.pdf
25. McConnell, S.: Code Complete. Pearson Education (2004)

26. Mokhov, S.A.: The use of machine learning with signal- and NLP processing of source code to fingerprint, detect, and classify vulnerabilities and weaknesses with MARFCAT. arXiv, Cryptography and Security (2011)
27. Mokhov, S.A., Paquet, J., Debbabi, M.: MARFCAT: fast code analysis for defects and vulnerabilities. In: 2015 IEEE 1st International Workshop on Software Analytics (SWAN), pp. 35–38 (2015). https://doi.org/10.1109/SWAN.2015.7070488
28. Narayanan, A., Chandramohan, M., Venkatesan, R., Chen, L., Liu, Y., Jaiswal, S.: Graph2vec: learning distributed representations of graphs. arXiv abs/1707.05005 (2017)
29. NIST: Software assurance reference dataset. https://samate.nist.gov/SRD/index.php. Accessed Mar 2022
30. NIST: National vulnerability database. https://nvd.nist.gov/. Accessed Nov 2021
31. RAT: rough-auditing-tool-for-security. https://code.google.com/archive/p/rough-auditing-tool-for-security/. Accessed May 2022
32. Reimers, N., Gurevych, I.: Reporting score distributions makes a difference: performance study of LSTM-networks for sequence tagging. In: Proceedings of the 2017 Conference on Empirical Methods in Natural Language Processing, Copenhagen, Denmark, pp. 338–348. Association for Computational Linguistics (2017). https://doi.org/10.18653/v1/D17-1035. https://aclanthology.org/D17-1035
33. Russell, R., et al.: Automated vulnerability detection in source code using deep representation learning. In: 2018 17th IEEE International Conference on Machine Learning and Applications (ICMLA), pp. 757–762 (2018). https://doi.org/10.1109/ICMLA.2018.00120
34. Schuster, M., Paliwal, K.: Bidirectional recurrent neural networks. IEEE Trans. Signal Process. **45**(11), 2673–2681 (1997). https://doi.org/10.1109/78.650093
35. Sestili, C.D., Snavely, W., VanHoudnos, N.M.: Towards security defect prediction with AI. arXiv abs/1808.09897 (2018)
36. Sharma, T., Kechagia, M., Georgiou, S., Tiwari, R., Sarro, F.: A survey on machine learning techniques for source code analysis. arXiv abs/2110.09610 (2021)
37. Shervashidze, N., Schweitzer, P., van Leeuwen, E.J., Mehlhorn, K., Borgwardt, K.M.: Weisfeiler-lehman graph kernels. J. Mach. Learn. Res. **12**(77), 2539–2561 (2011). http://jmlr.org/papers/v12/shervashidze11a.html
38. Wang, Z., Yu, L., Wang, S., Liu, P.: Spotting silent buffer overflows in execution trace through graph neural network assisted data flow analysis. arXiv (2021). https://arxiv.org/abs/2102.10452
39. Wanjia: This 66-year-old is still writing code and wants to fix bugs early in the SDLC. https://xcalibyte.com/. Accessed Mar 2022
40. Weiser, M.: Program slicing. IEEE Trans. Softw. Eng. **SE-10**(4), 352–357 (1984). https://doi.org/10.1109/TSE.1984.5010248
41. Yamaguchi, F., Golde, N., Arp, D., Rieck, K.: Modeling and discovering vulnerabilities with code property graphs. In: 2014 IEEE Symposium on Security and Privacy, pp. 590–604 (2014). https://doi.org/10.1109/SP.2014.44
42. Zhou, X., Verma, R.M.: Vulnerability detection via multimodal learning: datasets and analysis. In: ASIA Conference on Computer and Communications Security (2022). https://doi.org/10.1145/3488932.3527288

Assessing Deep Learning Predictions in Image-Based Malware Detection with Activation Maps

Giacomo Iadarola[1]([⊠])(iD), Francesco Mercaldo[1,2]([⊠])(iD), Fabio Martinelli[1],
and Antonella Santone[2]

[1] Institute of Informatics and Telematics (IIT),
National Research Council of Italy (CNR), Pisa, Italy
{giacomo.iadarola,francesco.mercaldo,fabio.martinelli}@iit.cnr.it
[2] Department of Medicine and Health Sciences "Vincenzo Tiberio",
University of Molise, Campobasso, Italy
{francesco.mercaldo,antonella.santone}@unimol.it

Abstract. Machine learning and deep learning models have been widely adopted to detect malware and protect our cyber infrastructures. The training is the most effective and important element of the artificial intelligence models. Nevertheless, it can be challenging and may require expertise and high-quality data. Inadequate training can be counterproductive, and lead to a model which may not detect the threats or, even worst, being exploited by the attackers. In this regard, the contribution of this short paper is twofold: we propose a method to (i) detect the malware belonging family and (i) provide reasoning about model evaluation and assess model soundness. The rationale behind this work aims to improve the evaluation of image-based deep learning models for malware family detection, especially in supervised learning tasks without recognizable or known patterns in the dataset samples. Our model obtains an overall accuracy of 0.934 in the evaluation of a dataset composed of 15726 real-world malware.

Keywords: Malware Analysis · Deep Learning · Explainable AI · Image-based Classifier · Cybersecurity

1 Introduction

Part of the reason for the perpetrated malware damages is due to the inadequacy of the commercial antimalware that use the signature-based paradigm; they provide automatic detection of malware by looking for malicious patterns (i.e. the signatures) stored in datasets of well-known threats. To overcome this limitation, researchers have been focusing on the adoption of machine learning, to change the signature-based paradigm and instead teaching models how to identify unknown malicious samples.

G. Lenzini and W. Meng (Eds.): STM 2022, LNCS 13867, pp. 104–114, 2023.
https://doi.org/10.1007/978-3-031-29504-1_6

In a nutshell, a machine learning model discovers and formalizes the principles and patterns shared by a set of data; with this knowledge, the algorithm can "reason" the properties of unedited samples. In malware detection, the data are the malicious and legitimate samples. Most of the malicious samples (the malware) are variants of an original sample, thus, they share part of the code and can be grouped in "families". Therefore, the machine learning model task is to discover similar or identical patterns between samples belonging to the same family, and exploit these patterns to classify newly discovered input samples.

In this short paper, we propose an approach for malware detection and family identification based on deep learning, with a particular focus on prediction assessment. Our approach exploits the representation of an application under analysis in terms of RGB images.

We compare two different deep networks architectures, the first one is a Convolutional Neural Network (CNN) designed by authors while the second one is a well-known CNN model for (generic) image classification task, the "Visual Geometry Group" (VGG16) model. Also, we visualise the so-called activation maps (i.e., the areas of the image symptomatic of a certain prediction) for the models obtaining the best performances in malware detection. This aspect allows us to understand the reason why the classifier output that specific prediction, and improve the explainability about the model decision.

The activation maps can help to understand why a certain sample is incorrectly classified and the differences between families. Nevertheless, understanding and exploiting the activation map information often requires prior knowledge of the malicious behaviour of the malware and expertise in the malware analysis field. Especially in the field of image-based malware analysis, where usually the target patterns are unknown due to the nature of the images, the model assessment plays an essential role but the evaluation of the model prediction robustness is a complex task. We try to overcome this problem by formalizing two novel evaluation metrics for evaluating activation maps in model assessment procedures, called *intra-family-SSIM* and *inter-models-SSIM*.

2   Background

To assess deep learning prediction and explainability in image-based malware detection, we exploits two Deep Learning (DL) models, a Grad-CAM model and the Structural Similarity Index Method (SSIM). In this section, a short overview on these techniques is reported but we refer to the literature for further details.

Image-Based Malware Detector. In a classical image classification task, such as identifying which animal is shown in a picture, DL models are able to correctly classify a sample because they are trained to recognize a specific pattern in the input image (i.e.: the pattern which represents the shape of that animal). Similarly, in the malware family classification, the application under analysis are classified by looking for the peculiar pattern of a specific malware family. Two malware belong to the same malware family if they share common malicious behaviour, thus, they share part of the source code (i.e.: the source code that performs the malicious behaviour). Indeed, the classical image classification and the malware family classification tasks are similar in their goal: identify a pattern able to discern an input sample into one of the output classes.

Therefore, converting malware to an image is the necessary but sufficient step to apply Deep Learning models for image classification tasks to malware family classification tasks. The conversion is straightforward, and widely adopted in literature [1,5,6]: every file stored in the hard disk of a computer can be represented in byte code, thus, we can cast each byte to an 8-bit unsigned integer (value in the range [0, 255]) which can be seen as a grayscale pixel. Similarly, we can group bytes and form pixels in the RGB colour model, to generate a three-channel colour information image. Figure 2 reports (on the left) a sample of malware converted to an image, belonging to the *Neoreklami* family.

Explainable Methods. Convolutional Neural Networks (CNNs) are widely adopted for Malware detection, and recently, researchers start studying techniques to explain the detection decisions. Explainable methods can be classified in different ways, but one of the most adopted classifications is the model-specific vs. model-agnostic techniques and local vs. global explanations. The model-specific techniques can interpret only the model for which they were devised; on the other hand, the model-agnosic techniques claim to be applied to any model. The local vs. global explanations regard the interpretability of the predictions of a model, the former explains a single output given a specific input, whereas the latter provides explanations of the whole model logic, without focusing on a specific input-output pair.

Most of the model-agnostic methods treat the model as a black box and analyse the output with regard to changes in the input. Local Interpretable Model Explanations (LIME) [7] is a widely adopted method for explaining predictions of models treated as a black box. LIME takes the input data sample, perturbs the data and analyse the impact on the output results. By doing so, it collects the set of "explanations" given by the input-output pairs and forms a local linear model that approximates the global model behaviour.

Our methodology is based on the explanations of single input-output pairs, thus, our approach lies in the local explanations techniques. Nevertheless, we also propose to group together the input-output pairs into output classes sets, to provide explanations for entire classes' decisions. We treat the model as a black box but our technique depends on the Convolutional layers, which makes the methodology (CNN-)model-specific.

Grad-CAM and SSIM. The Gradient-weighted Class Activation Mapping (Grad-CAM) is a technique to provide graphical information on the parts of the input image which have influenced the most the classification output of a CNN. The output of the Grad-CAM is a heatmap, that can be overlayed on the input image to highlight the relevant part. The heatmap is generated using the gradients of the final convolutional layers, which is the one that captures higher-level visual patterns and preserves spatial information on the input image.

The Grad-CAM adopted in this paper [9] is a relaxed generalization of the CAM approach [10]. The Grad-CAM does not require any modification to the CNN model architecture and provide a clear way to understand if the model has learnt correctly; that is, if the model is using the discriminative pattern in the input image to classify that sample. Intuitively, in an image-based malware classification task, the DL model should highlight the payload, (i.e.: the malicious code) of the malware, which is the shared pattern with the other malware of the family. Otherwise, if the model is classifying that sample because of another part of the input image than the payload, the malware could be easily modified by cutting out that highlighted part, preserving the malicious behaviour (express by the payload), and so generating a new malware variant, which will pass the model check as legitimate.

In contrast with the classical image classification task, in image-based malware classification, there are no recognizable shapes in the input image, at least for the human perception. Therefore, the heatmaps generated from the Grad-CAM do not have immediate usefulness because we are unaware of what constitutes the area they are highlighting. To compare the heatmaps and exploit their information, we use the Structure Similarity Index Method (SSIM), which is a perception based model to compare two different images [8]. Originally, the SSIM is used to evaluate image degradations, distortions and changes, such as contrast masking, in different versions of the same image. SSIM estimates the perceived quality of images by taking into account also spatially closed pixels and the different visibility (lightness) of the area in the different input image versions. The SSIM range values extend between -1 and $+1$ and only equal 1 if the two images are identical.

3   Methodology

The adoption of the methodology is not restricted to the DL model tested in our experiments (see Sect. 4), but it could be applied to any DL model capable of working with the Grad-CAM (i.e.: DL model with convolutional layers). Moreover, most of the process, excluding the dataset preprocessing, can be automatized and does not require prior knowledge on the malware or expertise on the topic. The methodology was implemented on a tool, freely available on GitHub [2].

For a better understanding of the methodology, it can be split into several steps, depicted in Fig. 1, which shows the approach we propose for assessing the prediction of a malware detection model built by means of deep learning. The first step of the methodology regards the collection of malware that compose the dataset. Every executable file can be taken into account, but the dataset has to be labelled and split into "Malware Families". One dataset class is reserved for "trusted" or "benign" applications. For the sake of simplicity, we refer generically to "malware families" all the classification output classes, even if one output is

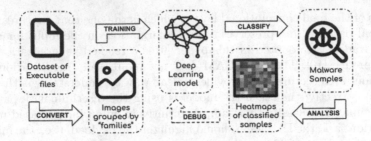

Fig. 1. Overall schema of the methodology.

reserved for trusted executables. Once the dataset is composed, the executables have to be converted into images. The images can be either in grayscale or colour format, but they should be similar in size. The DL model performs resizing on the input images, but if they differ too much in dimension the loss of information may compromise the DL model accuracy.

The images dataset is split into training, validation and test set, and a model assessment can be performed to find the best DL architecture and the best hyperparameters for the classification task. Moreover, and that is the key of the methodology, the model assessment takes into account not only the performance measures in the test (such as accuracy, precision, etc.) but also the "explainability" and robustness of the model by evaluating the Grad-CAM heatmaps. For instance, as reported in Sect. 4, we take into account two different CNN model architectures with similar accuracy, and we use the IF-SSIM and IM-SSIM for evaluating their "inference step", and ensure robustness in their outputs. The heatmap generated by the Grad-CAM can be overlayed on the input images (see Fig. 2), but the visual representation does not provide immediate and useful information; we can not easily understand what the highlighted area refer to in

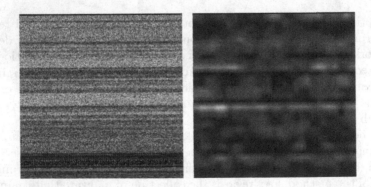

Fig. 2. Malware sample classified (correctly) as belonging to the *Neoreklami* family (on the left), next to the Grad-CAM output (the heatmap) of its evaluation (on the right).

the original source code of the application. Despite it is possible to reverse the process and reconstruct the original bytes of the application, as demonstrated in [3], there is the need of a soundness decompilation step, prior knowledge of the source code and time for performing the manual analysis.

Furthermore, in a previous work [4], we started exploring the use of the heatmaps for Image-based malware analysis, but we were not able to automatize or extract any information that would not require prior knowledge of the problem. Indeed, to ensure the robustness of the DL model, the adoption of heatmaps still would have required expertise to debug the deployed models. Recalling that the highlighted area should be the one covering the payload (the shared pattern between malware of the same family), different DL models trained on the same dataset should highlight the same area of the input image, otherwise, at least one of them is making a mistake. Exploiting this intuition, we propose to adopt the SSIM to evaluate differences between the heatmaps, and thus, the differences in the output predictions of the models. In detail, we propose two metrics called *inter-models-SSIM* IM-SSIM) and *intra-family-SSIM* (IF-SSIM), calculated as shown in Listing 1 and Listing 2, respectively, using the Python language. These metrics could be very useful in Image-based supervised learning tasks, when the pattern to identify in the input images is unknown (or not marked); for instance, in our experiments, we have a dataset of malware labelled by family but we are unable to locate the peculiar pattern of each family in their input samples.

The two novel metrics constitute the main core of the methodology, and they are explained in the next paragraphs in detail.

```
# Assuming that 'I' contains all the images classified by BOTH models
#  ('CNN_1' and 'CNN_2') belonging to the SAME malware family
def inter_models_SSIM(I, CNN_1, CNN_2):
    fam_avg = 0
    # Iterate over all images
    for i in I:
        # Generating heatmaps h_1 and h_2 from the same input i but
        #  different CNN models
        h_1 = grad_cam(CNN_1, i)
        h_2 = grad_cam(CNN_2, i)
        # adding the SSIM value to fam_avg
        fam_avg = fam_avg + SSIM(h_1,h_2)
    # calculate average SSIM of I where len(I) returns cardinality of I
    fam_avg = fam_avg / len(I)
    return fam_avg
```

Listing 1: Simplified python code to calculate the IM-SSIM metric for two DL models CNN_1 and CNN_2 and a given malware family.

IM-SSIM. This metric evaluates the performances of different trained models on the same test set by exploiting the differences in the heatmaps for the same malware family. Specifically, given a malware family and a couple of trained models, we take the set of input images that were classified into that family by both of the models. Then, we generate the two heatmaps for each input image, one for each model under analysis, and compare them with the SSIM metric. The SSIM values for each couple of heatmaps are summed up for all the input images classified into that malware family and the average is computed. The IM-SSIM value represents how much the two models differ in classifying the same samples for that specific malware family.

If the training is correct, the model will highlight in the input image the pattern that represents the malware payload, the peculiar pattern for that malware family and the only one that contains the malicious behaviour. Nevertheless, we do not have the information on the payload localization, thus, evaluating only the accuracy performance may lead to incomplete and misleading models. The IM-SSIM contributes to overcoming this problem by evaluating the model inference among a set of trained models. Intuitively, the heatmaps should be very similar across different models, because they should highlight the same pattern in the input images, and then IM-SSIM should tend to 1. If so, we can exploit this information and shrink the area to look for the family pattern. Otherwise, it means that (at least) one of the models (may also) correctly classify a sample but using an incorrect pattern, not the one peculiar to that malware family.

```python
# Assuming that 'I' contains all the images classified by model 'CNN_1'
#  belonging to the SAME malware family
def intra_family_SSIM(I, CNN_1):
  fam_avg = 0
  # map grad_cam(i,CNN_1) to all images 'i' in I and collect
  #  all the heatmaps in 'H'
  H = get_heatmaps(I, CNN_1)
  # generate all combinations of size 2 of H
  P = combinations(H, 2)
  # Iterate over all combinations of H
  for (h_1, h_2) in P:
    # adding SSIM values to single_avg
    fam_avg = fam_avg + SSIM(h_1, h_2)
  # calculate average SSIM of H, where len(H) returns cardinality of H
  fam_avg = fam_avg / len(H)
  return fam_avg
```

Listing 2: Simplified python code to calculate the IF-SSIM metric for a DL model CNN_1 and a given malware family.

IF-SSIM. The second novel metric focuses on a single model, and outputs information on how much the heatmaps of the same model and the same malware family differ from each other. While the IM-SSIM works on a subset of models, the IF-SSIM provides information on a single model, for a specific given malware family.

Once the model is trained, we take the set of input images classified into a specific family and generate the heatmaps. Then, we calculate the SSIM of each heatmap compared with any other, and we compute the average SSIM value. Finally, the IF-SSIM is computed as the average of these averages SSIM values. This algorithm requires many calculations, but it can be proved [2] that the result value is equivalent to calculating the average SSIM value of all the combinations of the list of heatmaps, which requires half of the operations; Listing 2 shows this more efficient approach.

These two metrics (i.e. the IM-SSIM and IF-SSIM) can be used in conjunction with the classical measure performance metrics (such as accuracy, precision, recall, F-measure and AUC) to provide a wider analysis of the DL models, also in case of unknown target patterns in the dataset samples. It is worth noting that both the metrics apply to the classified samples set, which also (may) contains misclassifications, and do not take into account the sample true label. Indeed, the metrics can be used in a real-world scenario with new and not labelled samples, because they evaluate the overall robustness of the model inference step, instead of single sample accuracy (unknown for unlabelled samples).

4 Experiments

Dataset. We collect 15726 malware samples split into 26 families (25 malware families and 1 trusted class). Each malware family count around 500 samples, except the *Ransom* and the *Trusted* ones which gather around 2000 samples each. Namely, all the malware family are: *Agent, Allaple, Amonetize, Androm, Autorun, BrowseFox, Dinwod, Elex, Expiro, Fasong, HackKMS, Hlux, Injector, InstallCore, MultiPlug, Neoreklami, Neshta, Ransom, Regrun, Sality, Snarasite, Stantinko, Trusted, VBKrypt,* and *Vilsel.* The DL models performance results (see Table 1) take into account the entire dataset, but due to the paper's limited length, we analyse only a couple of interesting families in detail (see Table 2). The GitHub repository [2] reports also the complete experimental results for all the families. The dataset samples were split into training, validation and test set, with a ratio of 80/10/10 respectively, and equally distributing the family samples.

Results and Discussion. We trained two different CNN models architecture, which we refer to with the name of CNN and VGG16. The main difference between them is the size of the architecture: the so-called CNN counts only 6 layers, instead of the 16 layers of the VGG16. The purpose of our experiments was to prove the usefulness of the new performance measures in the case of similar DL models. The complete DL architectures and experiments detail is reported in [2].

Table 1 reports the test results of the two DL models. Both models were trained on the same size input image (250 × 250 pixels in RGB colour). The results are comparable but the CNN performed slightly better than the VGG16.

Table 1. Comparison between the results on the test set.

DL model	CNN	VGG16
Img size x channels	250 x 3	
Epochs & Batch Size	20 & 32	
Layers	6	16
Loss	0.35	0.327
Accuracy	0.934	0.93
Precision	0.951	0.936
Recall	0.927	0.926
F-Measure	0.939	0.931
AUC	0.989	0.99

Table 2. Performance results on a subset of malware families.

	Model	Acc.	Prec.	Recall	F-Meas.	AUC	IF-SSIM	IM-SSIM
Hlux	CNN	1.000	1.000	1.000	1.000	1.000	0.925	0.333
	VGG16	0.996	0.893	1.000	0.943	0.998	0.952	
Neoreklami	CNN	0.998	0.980	0.960	0.970	0.980	0.649	0.493
	VGG16	0.999	1.000	0.980	0.990	0.990	0.668	
Ransom	CNN	0.994	0.966	0.985	0.975	0.990	0.247	0.272
	VGG16	0.987	0.950	0.945	0.947	0.969	0.452	
Sality	CNN	0.978	0.683	0.549	0.609	0.770	0.280	0.272
	VGG	0.980	0.711	0.627	0.667	0.810	0.478	
Trusted	CNN	0.969	0.838	0.897	0.866	0.937	0.285	0.280
	VGG16	0.966	0.817	0.897	0.855	0.936	0.470	
Vilsel	CNN	1.000	1.000	1.000	1.000	1.000	0.782	0.199
	VGG16	1.000	1.000	1.000	1.000	1.000	0.876	

On the other hand, the results in Table 2 shows that the models are not so robust as evaluating the standard measure performances only would express. The DL models on the *Hlux* family (as reported in Table 2) achieved very high accuracy values, and also very high IF-SSIM values, both of the models. Therefore, both of the models use one (and only) specific area of the input samples to classify (correctly) the malware; they use only the information in this limited input area to perform the classification since the IF-SSIM tend to 1, thus the heatmaps are all similar. Nevertheless, the IM-SSIM value of 0.333 means that the models do not agree: the input area used by one model is different from the area of the other. Similarly, it happens for the *Vilsel* family, that reports a lower IM-SSIM value of 0.199. Instead, the *Neoreklami* family appears to have a higher IM-SSIM and similar IF-SSIM values; the two models appear to have found a similar input area/pattern to classify these samples with very good accuracies.

One more interesting result is coming from the comparison between the *Ransom* and *Sality* families. The models have comparable IM-SSIM and IF-SSIM but different accuracy performances. Finally, the IF-SSIM help in evaluating the *Trusted* class, which is the only class that gathers executables that do not have a shared common pattern. In this case, the IF-SSIM value should be low because the heatmaps should differ. This can also be useful for assessing the quality and diversity of the *Trusted* samples.

5 Conclusion and Future Works

This paper proposes a CNN model for Image-based malware family classification. We tested the DL model on 15726 malware samples, split into 25 malware families and 1 Trusted class, and we achieved 0.934 accuracy on the test set. Also, we propose two novel performance metrics (*intra-family-SSIM* and *inter-models-SSIM*) to help the analysis of the DL performances; the metrics do not require prior knowledge on the dataset or the task, and they contribute to formalize and consider robustness and reliability in the classification process.

Despite the fact that the IF-SSIM and IM-SSIM can be used without any prior knowledge of the classification task, more experiments are needed to study the average values and improve the readability of these metrics, such as adding text explanation for given values ranges.

Also, the authors plan to extend this work by experimenting with more DL models and a larger dataset, and testing the usefulness of the metrics when evaluating adversarial samples.

Acknowledgment. This work has been partially supported by MIUR SecureOpen-Nets, EU SPARTA, CyberSANE and E-CORRIDOR projects. The authors also would like to thank Iacopo Ripoli for his help in the theoretical verification of the methodology.

References

1. He, K., Kim, D.S.: Malware detection with malware images using deep learning techniques. In: 2019 18th IEEE International Conference on Trust, Security and Privacy in Computing and Communications/13th IEEE International Conference on Big Data Science and Engineering (TrustCom/BigDataSE), pp. 95–102. IEEE (2019)
2. Iadarola, G.: Tool for image-based malware code analysis (2022). https://github.com/Djack1010/claransom. Accessed Oct 2022
3. Iadarola, G., Casolare, R., Martinelli, F., Mercaldo, F., Peluso, C., Santone, A.: A semi-automated explainability-driven approach for malware analysis through deep learning. In: 2021 International Joint Conference on Neural Networks (IJCNN), pp. 1–8. IEEE (2021)
4. Iadarola, G., Martinelli, F., Mercaldo, F., Santone, A.: Evaluating deep learning classification reliability in android malware family detection. In: 2020 IEEE International Symposium on Software Reliability Engineering Workshops (ISSREW), pp. 255–260. IEEE (2020)
5. Son, T.T., Lee, C., Le-Minh, H., Aslam, N., Raza, M., Long, N.Q.: An evaluation of image-based malware classification using machine learning. In: Hernes, M., Wojtkiewicz, K., Szczerbicki, E. (eds.) ICCCI 2020. CCIS, vol. 1287, pp. 125–138. Springer, Cham (2020). https://doi.org/10.1007/978-3-030-63119-2_11

6. Mercaldo, F., Santone, A.: Deep learning for image-based mobile malware detection. J. Comput. Virol. Hacking Tech. 1–15 (2020)
7. Ribeiro, M.T., Singh, S., Guestrin, C.: "Why should I trust you?": explaining the predictions of any classifier. In: Proceedings of the 22nd ACM SIGKDD International Conference on Knowledge Discovery and Data Mining, San Francisco, CA, USA, 13–17 August 2016, pp. 1135–1144 (2016)
8. Sara, U., Akter, M., Uddin, M.S.: Image quality assessment through FSIM, SSIM, MSE and PSNR-a comparative study. J. Comput. Commun. **7**(3), 8–18 (2019)
9. Selvaraju, R.R., Cogswell, M., Das, A., Vedantam, R., Parikh, D., Batra, D.: Grad-CAM: visual explanations from deep networks via gradient-based localization. In: Proceedings of the IEEE International Conference on Computer Vision, pp. 618–626 (2017)
10. Zhou, B., Khosla, A., Lapedriza, A., Oliva, A., Torralba, A.: Learning deep features for discriminative localization. In: Proceedings of the IEEE Conference on Computer Vision and Pattern Recognition, pp. 2921–2929 (2016)

Data Analysis of Security and Trust

M²M: A General Method to Perform Various Data Analysis Tasks from a Differentially Private Sketch

Florimond Houssiau[1(✉)], Vincent Schellekens[2], Antoine Chatalic[3], Shreyas Kumar Annamraju[4], and Yves-Alexandre de Montjoye[4]

[1] The Alan Turing Institute, London, UK
fhoussiau@turing.ac.uk
[2] UCLouvain, Ottignies-Louvain-la-Neuve, Belgium
[3] MaLGa - DIBRIS, Università di Genova, Genoa, Italy
[4] Imperial College London, London, UK
demontjoye@imperial.ac.uk

Abstract. Differential privacy is the standard privacy definition for performing analyses over sensitive data. Yet, its privacy budget bounds the number of tasks an analyst can perform with reasonable accuracy, which makes it challenging to deploy in practice. This can be alleviated by private sketching, where the dataset is compressed into a single noisy sketch vector which can be shared with the analysts and used to perform arbitrarily many analyses. However, the algorithms to perform specific tasks from sketches must be developed on a case-by-case basis, which is a major impediment to their use. In this paper, we introduce the generic *moment-to-moment* (M²M) method to perform a wide range of data exploration tasks from a single private sketch. Among other things, this method can be used to estimate empirical moments of attributes, the covariance matrix, counting queries (including histograms), and regression models. Our method treats the sketching mechanism as a black-box operation, and can thus be applied to a wide variety of sketches from the literature, widening their ranges of applications without further engineering or privacy loss, and removing some of the technical barriers to the wider adoption of sketches for data exploration under differential privacy. We validate our method with data exploration tasks on artificial and real-world data, and show that it can be used to reliably estimate statistics and train classification models from private sketches.

Keywords: Privacy · Differential privacy · Sketching · Sketched learning

1 Introduction

The amount and level of detail of data collected has increased exponentially over the last two decades. Behavioral data has evolved from hand-collected medical

---

F. Houssiau and V. Schellekens—These authors contributed equally.

G. Lenzini and W. Meng (Eds.): STM 2022, LNCS 13867, pp. 117–135, 2023.
https://doi.org/10.1007/978-3-031-29504-1_7

records to GPS traces automatically recorded with a temporal resolution on the scale of seconds. While this increased availability and precision of data has resulted in tremendous advances in research, they raise serious privacy concerns. Modern datasets often contain highly detailed summaries of our lives, and are notoriously hard to anonymize. Individuals have indeed been shown to be easily re-identifiable in large-scale behavioral datasets, such as mobile phone meta-data [27], credit card data [28] and web browsing data [5].

Differential privacy (DP) [14] was introduced by Dwork et al. as a property of algorithms that protect the privacy of users in a dataset. It requires for a randomized algorithm's outputs to be distributed approximately identically whether any one individual is in the dataset or not. The discrepancy between distributions is controlled by a parameter ε known as the privacy budget. DP is considered by many to as the gold standard definition for privacy loss in aggregated data releases. DP mechanisms have been deployed by institutions with access to large datasets, such as Google to measure changes in mobility patterns caused by confinement measures [2], LinkedIn to answer analytics queries [21], and the US Census for the 2020 Census [1].

Most applications of DP remain limited to specialized tasks on large datasets. Indeed, each differentially private access to a dataset consumes some privacy budget ε, and the total acceptable budget is fixed by the data owner for the dataset. Once this budget has been used entirely, the dataset must be discarded. As such, the number of accurate statistical tasks an analyst can run on a dataset is capped. This strongly limits the utility of differential privacy in practice. In particular, data exploration with DP is particularly challenging: it requires analysts to establish which analyses they want to perform on the dataset, and how to divide the budget between them, before accessing the data.

An increasingly popular solution to this issue is to first compute a differentially private summary of the data, called a private *sketch*, which is then shared with analysts. Once computed, the private sketch can be used as much as desired to solve new learning tasks without accessing the data anymore or using additional privacy budget. This follows from the post-processing property of DP. Sketches have long been used as a technique to compress large-scale datasets to reduce the computational load of algorithms. In this work, the sketch of a dataset is defined as the empirical average of some *feature map* function Φ over all records in a dataset D: $z_D = \frac{1}{|D|} \sum_{x_i \in D} \Phi(x_i)$. The choice of feature map controls the specificity of the information contained in the sketch. For example, researchers have proposed sketches based on Random Fourier Features (RFF) [35] and locality-sensitive hashing [11] that approximate kernel density estimates of the empirical distribution. For some specific sketches and tasks, algorithms with strong theoretical guarantees of accuracy have been developed [17].

However, performing arbitrary data analysis tasks from sketches is difficult, as extracting the desired information from a highly compressed representation of the data is challenging. Each specific task and feature map Φ would require a dedicated algorithm designed by experts. For instance, RFF sketches have in practice only been used for a few tasks, such as Gaussian mixture modeling (GMM) [22] or k-means [23]. Developing compressive methods for other data

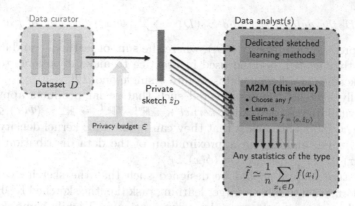

Fig. 1. Considered setup. The data curator releases "once and for all" a private sketch with privacy budget ε. The analyst then chooses a function f and uses our M²M method to learn a vector $a \in \mathbb{R}^m$ such that $\widetilde{f} = \langle a, \hat{z}_D \rangle$ approximates the empirical average value of f over the dataset, $\overline{f} = \frac{1}{n} \sum_i f(x_i)$. This procedure can be repeated any number of times (for various choices of f) without additional privacy budget.

exploration tasks remains an open problem. This is the main obstacle to using sketches for general data analysis.

In this paper, we introduce a **heuristic to learn from dataset sketches** as shown in Fig. 1, which we call the moment-to-moment (M²M) method. M²M allows to approximate empirical averages of functions f from the sketch, $\frac{1}{|D|} \sum_{x_i \in D} f(x_i)$, and can in principle be applied to any feature map Φ. This method is inspired by approximation techniques for kernel methods using random features [25,32]. We **empirically validate our method with artificial and real-world data**, and show that a variety of tasks (moment estimation, counting queries, covariance estimation, logistic regression) can be learned from sketches with comparable performances to alternatives (synthetic data).

2 Background

2.1 Sketches

Sketches are compressed representations of data collections, which can be used to perform some operations efficiently but approximately [6,12]. Sketches usually rely on randomness to achieve a compact representation size. This comes at the price of a probabilistic approximation error. This general principle finds applications in a broad set of contexts, from data streams [16,26] to randomized linear algebra [13]. Here, we focus on sketches that compress the dataset $D = (x_1, \ldots, x_n)$ to a single sketch vector $z_D \in \mathbb{R}^m$ by computing the *average of a nonlinear feature map* Φ, applied to each record x_i.

Definition 1. *Given a feature map* $\Phi : \mathbb{R}^d \to \mathbb{R}^m$, *the* sketch *of a dataset* $D = (x_1, \ldots, x_n) \in \mathcal{D}$ *is*

$$z_D \triangleq \frac{\Sigma_\Phi(D)}{|D|} = \frac{1}{n} \sum_{i=1}^n \Phi(x_i) \in \mathbb{R}^m, \tag{1}$$

with $n = |D|$ the dataset size and $\Sigma_\Phi(D) = \sum_{i=1}^{n} \Phi(x_i)$ the sum of features.

The representation $(\Sigma_\Phi(D), |D|)$, where the sum-of-features and dataset size are distinctly encoded, is often used in practice to make it possible to further combine sketches of different datasets into a single one [12].

Typically, sketches are constructed such that scalar products approximate a specific similarity score (called kernel $\kappa : \mathbb{R}^d \times \mathbb{R}^d \rightarrow \mathbb{R}_+$), $\langle \Phi(x), \Phi(x') \rangle \simeq \kappa(x, x') \; \forall x, x'$ [30]. This means that they can be used for kernel density estimation (KDE) , i.e. building an approximation of the data distribution p_X by a density $\hat{p}(x) \triangleq \frac{1}{n} \sum_{i=1}^{n} \kappa(x, x_i) \approx \langle \phi(x), z_D \rangle$.

The feature map Φ should be designed such that the sketch z_D captures enough information to solve a target learning task (i.e. the sketched KDE density $\tilde{p}$ accurately represents the true data distribution p_X) while compressing the data as much as possible (i.e. the sketch size m should be small). We here review several important feature map choices.

Histograms. Histograms and contingency tables have been extensively studied in the DP literature [14]. Both can be seen as a illustrative examples of sketches (in the sense of Definition 1), where the feature map is

$$\Phi^{\mathrm{HIST}}(x) \triangleq (I\{x \in \mathcal{P}_i\})_{i=1,\ldots,m} \in \{0,1\}^m,$$

where $(\mathcal{P}_i)_{i=1}^{m}$ is a list of subsets of the data domain $\mathbb{R}^d$, and $I\{A\}$ is the indicator function which returns 1 (resp. 0) whenever A is true (resp. false). For 1-D histograms with n_{bins} bins for example (what we call the HIST sketch), these sets are the one-dimensional bins along each component. For this sketch, $m = d \cdot n_{bins}$.

RFF Sketches. Random Fourier Features (RFF) aim to approximate shift-invariant kernels $\kappa(x, x') = K(x - x')$. They were initially introduced to accelerate kernel methods in machine learning [32].

Definition 2 (Random Fourier Features). Given $m' = \frac{m}{2}$ "frequency vectors" $\Omega = [\omega_1, \ldots, \omega_{m'}] \in \mathbb{R}^{d \times m'}$ drawn $\omega_j \sim_{i.i.d.} \Lambda$, the RFF map is defined as:

$$\Phi^{\mathrm{RFF}}(x) \triangleq [\cos(x^T \Omega), \sin(x^T \Omega)]^T \in \mathbb{R}^m.$$

The idea is that shift-invariant kernels can be decomposed as $K(x - x') = \mathbb{E}_{\omega \sim \Lambda} e^{i \omega^T (x - x')}$ where the probability distribution Λ is the kernel Fourier transform $\Lambda(\omega) = \int K(u) e^{-i \omega^T u} du$ (owing to Bochner's theorem [34]). For example, the Gaussian kernel $\kappa(x, x') = \exp\left(-\|x - x'\|_2^2 / 2\sigma^2\right)$ admits a Gaussian distribution as Fourier transform, $\Lambda = \mathcal{N}(0, \sigma^{-2} I_d)$. One can then show [32] that up to a constant scaling, Φ^{RFF} satisfies the kernel equation for this kernel.

RFF sketches have been successfully used for parametric density estimation tasks, such as $k-$means [23] and Gaussian Mixture Modeling [22], reducing the computational resources required by orders of magnitude on large-scale datasets.

RACE Sketches. The Repeated Array-of-Counts Estimator (RACE) sketch was proposed [11] as an alternative way to approximate KDE for so-called LSH kernels. In RACE, the feature map Φ takes binary values, and is constructed by concatenating R independent hashing functions that each map to W distinct buckets. The size of the sketch is thus $m = R \cdot W$. RACE sketches use *locally-sensitive* hash (LSH) functions: let $W \in \mathbb{N}_0$, a family $\mathcal{H}$ of hash functions $h : \mathbb{R}^d \rightarrow \{1, ..., W\}$ is locally-sensitive with collision probability κ if $\mathbb{P}_{\mathcal{H}}[h(x) = h(x')] = \kappa(x, x')$ for all $x, x' \in \mathbb{R}^d$.

Definition 3 (Repeated Array-of-Counts Estimator). *Given $W \in \mathbb{N}_0$, $h_j, j = 1, ..., R$, a set of $R = \frac{m}{W}$ hash functions drawn independently from $\mathcal{H}$, the associated RACE map is defined as:*

$$\Phi^{\mathrm{RACE}}(x) \triangleq \left[\iota(h_1(x))^T, ..., \iota(h_R(x))^T \right]^T \in \{0, 1\}^m,$$

where $\iota : \{1, ..., W\} \rightarrow \{0, 1\}^W$ denotes the one-hot encoding operation.

Similarly to RFF, one can show [11] that for all choices of LSH, there exists a kernel κ such that the kernel equation is satisfied.

2.2 Differential Privacy

Differential privacy (DP) [14] is seen as the standard definition of privacy for aggregate data releases. It states that the distribution of a differentially private algorithm's output is similar for any two *neighboring* datasets. Different relations can be considered, but in general (and for the rest of this manuscript), we consider that two datasets are neighbors if they differ by the addition or removal of any one record; this is known as "unbounded" DP[1]. The guarantees of DP are characterized by a privacy "budget" $\varepsilon > 0$ which bounds the information disclosure from the dataset. Denote by $\mathcal{D}$ the set of all datasets, equipped with a neighboring relation $\sim$. In this work, we consider datasets as collections of d-dimensional real-valued vectors $x_i \in \mathbb{R}^d$.

Definition 4 (Differential Privacy). *A randomized mechanism $\mathcal{M} : \mathcal{D} \rightarrow \mathbb{R}^m$ is ε-differentially private iff $\forall D \sim D' \in \mathcal{D}$, $\forall S \subset \mathbb{R}^m$:*

$$\mathbb{P}[\mathcal{M}(D) \in S] \leq e^\varepsilon \, \mathbb{P}[\mathcal{M}(D') \in S].$$

Differential privacy has several desirable properties. First, *composition* guarantees that accessing the same dataset with N different mechanisms respectively using budgets $\varepsilon_1, ..., \varepsilon_N$ uses a total budget of at most $\varepsilon_{total} = \sum_{i=1}^N \varepsilon_i$. Second, *post-processing* ensures that once some quantities have been computed by a differentially private algorithm, no further operation on these quantities can

---

[1] We consider only unbounded DP for conciseness, yet the private sketches from Sect. 2.3 can be extended in a straightforward manner to the bounded DP setting. In this case no noise needs to be added to the denominator in (2).

weaken the privacy guarantees. The latter is particularly important for sketches, as it implies that all analyses ran on a ε−DP sketch are ε−differentially private.

A common method to compute a function f over a dataset with ε-DP is the Laplace mechanism [15]. For a target function $f : \mathcal{D} \to \mathbb{R}^m$, this mechanism adds centered Laplace noise with scale proportional to the *sensitivity* of f.

Definition 5 (Laplace Mechanism). *The Laplace mechanism to estimate privately a function* $f : \mathcal{D} \to \mathbb{R}^m$ *is defined as* $\mathcal{M}_f^{\mathcal{L}}(D) = f(D) + \xi$, *where* $\xi_j \sim \mathcal{L}(\beta), j = 1, ..., m$ *is centered Laplace noise with scale parameter* $\beta = \frac{\Delta_1(f)}{\varepsilon}$. *The sensitivity* $\Delta_1(f)$ *is defined as* $\Delta_1(f) \triangleq \sup_{D \sim D'} \|f(D) - f(D')\|_1$.

2.3 Differentially Private Sketching

We new consider privatized versions of the sketches in the form (1). As the considered feature maps are bounded, their sensitivities are also easily bounded; thus, the Laplace mechanism can be used to produce private versions of these sketches. Following [8] we compute a sketch of the form

$$\hat{z}_D \triangleq \frac{\Sigma_\Phi(D) + \xi}{|D| + \zeta} \triangleq \frac{\sum_{i=1}^n \Phi(x_i) + \xi}{n + \zeta}, \tag{2}$$

where ξ_j $(j = 1, ..., m)$ and ζ are all Laplace random variables with scale parameter chosen according to Definition 5. For ξ, the scale depends on the sensitivity of the sum-of-features function, which can be expressed as $\Delta_1(\Sigma_\Phi) = \max_x \|\Phi(x)\|_1$, which can be computed as: $m'\sqrt{2}$ for RFF [8], R for RACE [11], and k for HIST [14]. For ζ the scale parameter depends on the sensitivity of the cardinality function which is always $\Delta_1(|\cdot|) = 1$. The total privacy budget ε is split across the numerator and the denominator, i.e. the noises ξ and ζ are also respectively proportional to ε_{num}^{-1} and ε_{den}^{-1}, with $\varepsilon = \varepsilon_{num} + \varepsilon_{den}$. As stated above, such private sketches have already been considered in the literature and are not a contribution of this paper: we simply use sketches of this form in order to apply the M²M method introduced in the next section.

Although we focus on pure ε-DP in this manuscript for simplicity, private sketches can easily be extended to satisfy approximate DP (also known as (ε, δ)−differential privacy) using the Gaussian mechanism [15]. This requires computing the L_2 sensitivity of the feature map, see for example [8,18] for RFF.

2.4 Related Work

The key advantages of sketching methods for data analysis with differential privacy is that they produce a private "summary" of the dataset, from which an arbitrary number of analyses can be performed. This idea of publishing a DP summary of the data has been explored in the literature, e.g., by Barak et al. with the release of full contingency tables [4]. As contingency tables do not scale with the number of dimensions, further work has been proposed to publish so-called "views" of the data, from which n−way marginals can then be computed [31]. Another type of data summary that has gained popularity in recent years is *synthetic data*, where the data curator publishes a dataset is "similar" to the original data, but with no mapping from real to synthetic records. These usually involve training a statistical model on the data, which is then used to generate synthetic records, either explicitly [24,37] or using generative networks [36].

Kernel mean embeddings are known to carry a lot of information on the data distribution and are thus of particular interest for privacy applications. Balog et al. suggested to use synthetic data points in order to represent (possibly infinite-dimensional) kernel mean embeddings in a private manner [3]. Finite-dimensional approximations based on random Fourier features have been made private using simple additive perturbation mechanisms with applications to clustering and Gaussian modeling [8] as well as synthetic data generation [18]. More recently, compact sketches based on Hermite polynomials have been proposed [29] and have been shown empirically to provide a better privacy-utility tradeoff for private data generation than random Fourier features.

Relating specifically to the M²M method, the idea of considering a learned linear combination of random features (without privacy) has been popularized by Rahimi and Recht [33], and then extensively studied under the name of "extreme learning machines" (ELMs) [19,20]. The sketches considered in this paper can be interpreted as instances of this idea, with an additional averaging operation over the dataset. This is made possible by the fact that we only consider learning moments of the data.

3   The Moment-to-Moment Method

Sketching methods can be used to efficiently perform specific learning tasks, and can often be made private in a straightforward manner by additive perturbation; however, extracting information from them is hard in general. Here, we introduce the *moment-to-moment* (M²M) heuristic to learn a broad range of aggregate statistics from a single sketch. While previous sketched learning methods were relatively specific in the sense that both the feature map and the algorithm to learn from the sketch had to be tailored to a specific machine learning task, our heuristic can be used to approximate various kinds of statistics from the same sketch. Although M²M can naturally be used on a non-private sketch, it is particularly attractive for private sketches, as it allows an analyst to perform arbitrarily many analyses from the sketches without incurring any additional privacy budget.

In the following, we assume that the data curator holds a sensitive dataset D of size n, chooses a data-independent feature map Φ, and releases publicly the triplet $(\Phi, \hat{z}_D, n + \zeta)$ where $\hat{z}_D = \frac{1}{n+\zeta}(\sum_{i=1}^{n} \Phi(x_i) + \xi)$ is the private sketch computed as in (2) and ξ, ζ are random and chosen as explained in Sect. 2.3 in order to ensure ε-DP (i.e. they depend on the sensitivity of the feature map Φ). Note that any result obtained by post-processing from this triplet will always remain ε−DP.

3.1   Method Description

Suppose that an analyst wants to compute the empirical average of an arbitrary target function $f : \mathbb{R}^d \to \mathbb{R}$ over the dataset, i.e. the quantity $\overline{f} \triangleq \frac{1}{n} \sum_{i=1}^{n} f(x_i)$. The M^2M method estimates $\overline{f}$ by a linear function of the sketch $\langle a, \hat{z}_D \rangle$, where the coefficients $a \in \mathbb{R}^m$ are computed by the method. Because both the input (the sketch $\hat{z}_D$) and the output (the target average $\overline{f}$) can be seen as "generalized moments" (averages of some features of the data) of the dataset, this amounts to transforming one type of generalized moment to another, hence the name of our method.

In order to apply the method, an analyst chooses *a priori* a bounded domain $D_{M^2M} \subset \mathbb{R}^d$ such that all possible records lie inside of D_{M^2M} (for example, D_{M^2M} might be a box constrained by physical upper and lower bounds on the data values). The principle of M^2M is to approximate the target function $f : \mathbb{R}^d \to \mathbb{R}$ over this domain D_{M^2M} by a linear model $\widetilde{f}$ of parameters $a \in \mathbb{R}^m$ in the output space of the sketch feature map $\Phi : \mathbb{R}^d \to \mathbb{R}^m$, i.e.,

$$\widetilde{f}(x) \triangleq \langle a, \Phi(x) \rangle \approx f(x), \quad \forall x \in D_{M^2M}.$$

The key insight is that this linear model can then be used to estimate the dataset average $\overline{f}$ from the dataset sketch z_D, since the sketching operator is linear:

$$\widetilde{f}(z_D) = \langle a, z_D \rangle = \frac{1}{n} \sum_{i=1}^{n} \langle a, \Phi(x_i) \rangle \approx \frac{1}{n} \sum_{i=1}^{n} f(x_i) = \overline{f}. \tag{3}$$

Intuitively, the target function $f(\cdot)$ is approximated by a linear combination $\langle a, \Phi(\cdot) \rangle$ of a set of m base functions (the components of the feature map, $\Phi_i(\cdot)$). The quality of the approximation depends on the compatibility between the feature map Φ and the function to approximate f. Zhang et al. [38] showed that such a linear combination can approximate continuous functions arbitrary well for a large enough number of features m, under conditions satisfied by many standard feature maps. This suggests that M^2M can be used to approximate any continuous function f for a large array of sketches, although quantifying precisely how the approximation quality decreases with m is out of the scope of this paper. It should also be expected that approximating a discontinuous function f with, e.g., RFF features will lead to high approximation error (e.g., some kind of Gibbs phenomenon).

We illustrate M^2M with a toy example in Fig. 2. We consider the step function $f(x) = I\{x \geq 0.5\}$ restricted to the domain $D_{M^2M} = [0,1] \subset \mathbb{R}^{d=1}$. For RFF, the approximation $\widetilde{f}(x)$ is a linear combination of $\cos(\omega_i^T x)$ and $\sin(\omega_i^T x)$, for some fixed ω_i, which explains the bumps observed in the approximation (Gibbs phenomenon). The RACE feature map, whose base functions are the one-hot encoding of locally-sensitive hash functions, approximates f by a piecewise constant function.

3.2   Optimizing the M^2M Model

For the results produced by the M^2M method to be useful, the parameters of the linear model a need to be chosen such that $\widetilde{f}(\cdot)$ is a good approximation of the

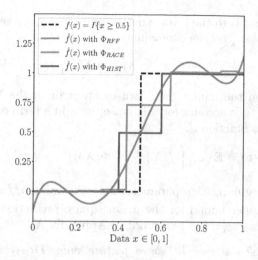

Fig. 2. How M²M works: the M²M method approximates the target function f as a linear combination of components of the feature map, $\sum_{i=1}^{m} a_i \Phi_i(x) \approx f(x)$.

true function $f(\cdot)$ on the domain of interest. For this, we formulate and optimize a loss function J for the vector of weights a that penalizes differences between f and $\tilde{f}$. The full learning procedure is described in Algorithm 1 in Appendix B.

Similarly to Rahimi et al. [32], we use the squared difference as distance, $d(f(x), \tilde{f}(x)) = (\tilde{f}(x) - f(x))^2$. Assume that the records $x_i \in D$ are drawn from some (unknown) probability distribution $x_i \sim_{i.i.d.} p_X$. Ideally, the M²M procedure would minimize the average error of the approximation *over the true data distribution* p_X, $J_{\text{ideal}}(a) = \mathbb{E}_{X \sim p_X}\left[d\left(f(x), \tilde{f}(x)\right)^2\right]$. However, the analyst only has access to the private sketch and does not know p_X, let alone the data D. Instead, they choose an *a priori* distribution ψ that is either (1) close to p_X, or (2) likely to yield a good approximation where p_X takes significant values when optimizing for the (approximate) loss $J_\psi(a) = \mathbb{E}_{X \sim \psi}\left[d\left(f(x), \tilde{f}(x)\right)^2\right]$. In this work, we assume no prior knowledge except for the domain $D_{\text{M²M}}$ and thus use the uniform distribution on this domain $\psi = Unif(D_{\text{M²M}})$, following the principle of maximum entropy. Finally, since evaluating the expectation operator analytically can be challenging for arbitrary ψ, f and Φ, especially in high dimensions, we approximate it by sampling a large number n_s of *training synthetic* data points $(\tilde{x}_i)_{i=1}^{n_s}$ sampled i.i.d. from ψ. The resulting loss, given a choice of ψ, is:

$$J_{\text{noreg}}(a) = \tfrac{1}{N}\sum_{i=1}^{N} (f(X_i) - \langle a, \Phi(X_i)\rangle)^2 \quad X_1, \ldots, X_N \sim_{\text{i.i.d.}} \psi$$

However, minimizing J_{noreg} directly is not robust to noise, and in particular to the noise added to obtain differential privacy. Indeed, when applying the linear model a from (3) to the private data summary $\hat{z}_D$, we get (neglecting, for illustration, the noise ζ on the denominator):

$$\langle \hat{z}_D, a \rangle = \langle \tfrac{1}{n} \textstyle\sum_{i=1}^n \Phi(x_i) + \xi, a \rangle \approx \tfrac{1}{n} \textstyle\sum_{i=1}^n f(x_i) + \tfrac{1}{n}\langle \xi, a \rangle.$$

Hence, the noise on the numerator ξ causes an error in the M^{2}M estimate of variance $\sigma_\xi^2 \|a\|_2^2 / n^2$. To account for this noise, we add a term proportional to its variance to the loss function J:

$$J(a) \triangleq \mathbb{E}_{X \sim \psi} \left[(f(X) - \langle a, \Phi(X) \rangle)^2 \right] + \lambda \|a\|_2^2, \tag{4}$$

where we set the regularization parameter λ to the value σ_ξ^2 / n^2. We prove that this loss J is an upper bound for the mean square prediction error between $\bar{f}$ and the M^{2}M estimate $\langle a, \hat{z}_D \rangle$ (see proof in Appendix A).

Theorem 1. *Let* $\Phi : \mathbb{R}^d \to \mathbb{R}^m$ *be a feature map,* $D_{M^2M} \subset \mathbb{R}^d$, *and* D *be a random dataset of n records* $X_1, \ldots, X_n \sim_{i.i.d.} \psi$. *For all* $a \in \mathbb{R}^m$, *and all distributions* ψ, *if* $\lambda = \sigma_\xi^2 / n^2$ *and* $\zeta = 0$, *we have that, if* $\zeta = 0$:

$$J(a) \geq \mathbb{E}_{X_1, \ldots, X_n, \xi} \left[\left(\tfrac{1}{n} \textstyle\sum_{i=1}^n f(X_i) - \langle a, \hat{z}_D \rangle \right)^2 \right]$$

Since the exact dataset size n is not directly available to the analyst, we use $|D| + \zeta$ as an estimation of n. Further to this, we found empirically that using $\lambda = \frac{\sigma_\xi^2}{n^2}$ makes the model insufficiently robust to noise (especially when the sensitivity of the feature map is large). We thus use a larger regularization parameter in experiments by removing the square on the estimated number of samples.

$$\lambda = \frac{\sigma_\xi^2}{(|D| + \zeta)} = \frac{2 \cdot \Delta_1(\Phi)^2}{\varepsilon_{num}^2 \cdot (|D| + \zeta)}. \tag{5}$$

Solving for J. Let $(\tilde{x}_i)_{i=1}^{n_s}$ denote the set of random training samples used inside the M^{2}M procedure. Denote the synthetic feature matrix $\mathbf{P} = (\Phi(\tilde{x}_i))_{i=1}^{n_s} \in \mathbb{R}^{n_s \times m}$, and the vector of corresponding outputs $\mathbf{F} = (f(\tilde{x}_i))_{i=1}^{n_s} \in \mathbb{R}^n$. The empirical loss that M^{2}M optimizes is $J(a) = \frac{1}{n_s} \|\mathbf{P} \cdot a - \mathbf{F}\|_2^2 + \lambda \|a\|_2^2$. This corresponds to a ridge regression problem with regularization parameter λ, and can be solved efficiently.

3.3 Sources of Error

M^{2}M is a heuristic method to approximate $\bar{f}$, and as such will always incur some error. We here outline the four main sources of error of M^{2}M.

1. *Sampling error*: The expectation operator in the cost function $J(a)$ is not computed exactly, but estimated by sampling n_s points $\tilde{x}_i \sim \psi$. If n_s is too small, this estimate can be inaccurate, and the model a risks "overfitting" to the small training set.

2. *Approximation error*: M²M finds coefficients a such that the linear combination $\tilde{f}(\cdot) = \langle a, \Phi(\cdot) \rangle = \sum_{i=1}^{m} a_i \Phi_i(\cdot)$ approximates the target function f. In general, even if a is the exact minimizer of $J(a)$, there remains some inherent approximation error which depends on the compatibility between the feature map Φ and target function f.

3. *Distributional shift*: In practice, the empirical distribution p_X differs from the probability distribution ψ used for training. Distributional shift is a hard problem to fix, as it requires tailoring ψ to p_X *without accessing the data*, or only through the sketch. We discuss this in Sect. 5.

4. *Differential privacy noise*: Finally, the noises ξ and ζ added in the computation of the sketch $\hat{z}_D$ further distort the representation. This error decreases when the privacy budget ε increases.

3.4 Statistical Estimation with M²M

Many learning tasks can be written as the estimation of some generalized moments of the data. Here we give some common examples.

1. Moments: The jth component of the kth moment of the empirical data distribution is defined as

$$ m_j^{(k)} = \tfrac{1}{n} \sum_{i=1}^{n} (x_i)_j^k \approx \mathbb{E}_{X \sim p_X} X_j^k, $$

which is the empirical average of the function $f^{(j,k)} : \mathbb{R}^d \to \mathbb{R} : x \mapsto x_j^k$.

2. Counting queries: Given a set $S \subset \mathbb{R}^d$, a counting query over D consists of finding the number of data points from the dataset D that belong to S:

$$ \text{COUNT}(D, S) = |\{i \in \{1, \ldots, n\} : D_i \in S\}| = \sum_{1 \le i \le n} f_S(x_i). $$

where $f_S : \mathbb{R}^d \to \{0, 1\} : x \mapsto I\{x \in S\}$ denotes the indicator function of S. Histograms are a specific subset of counting queries, where the set S is chosen to be a one-dimensional "bin".

3. Covariance: The $(i, j)^{\text{th}}$ entry of the empirical covariance matrix is

$$ c_{ij} = \tfrac{1}{n} \sum_{l=1}^{n} ((x_l)_i - \mu_i) \cdot ((x_l)_j - \mu_j), $$

which is the empirical average of the function $f^{(i,j)} : \mathbb{R}^d \to \mathbb{R} : x \mapsto (x_i - \mu_i)(x_j - \mu_j)$. The mean of the component i, μ_i, can be estimated using M²M for the first-order moment, $m_i^{(1)}$.

3.5 Classification and Regression by Approximation of the Loss

Many learning tasks can be formulated as learning a parametric model with parameter θ using a loss function L. For such tasks, one will typically solve the optimization problem $\theta^* \in \arg\min_\theta \tfrac{1}{n} \sum_{i=1}^{n} L(x_i, \theta)$, whose objective function takes the form of a generalized moment. Specifically, for a classification or regression task, the analyst wants to fit some model $F_\theta : \mathbb{R}^{d-1} \to \mathbb{R}$ parameterized by $\theta \in \mathbb{R}^p$ to the data samples $(x_i)_{i=1}^n$, where each sample x_i is a pair

$x_i = (\overline{x}_i \in \mathbb{R}^{d-1}, y_i \in \mathbb{R})$. If the fitting quality is quantified by a loss function $l(.,.)$, one can define $L_\theta(x_i) \triangleq L(x_i, \theta) \triangleq l\left(F_\theta(\overline{x}_i), y_i\right)$ and M^{2}M can be used with the target $f = L_\theta$ for any *fixed* value of θ. Finding the optimal parameter θ^* involves solving the following bi-level optimization problem:

$$\theta^* \in \arg\min_\theta \langle a_\theta, \hat{z}_D \rangle \quad \text{such that} \quad a_\theta \in \arg\min_a J_\theta(a) \tag{6}$$

where J_θ is the M^{2}M objective associated to the target function L_θ. As mentioned in Sect. 3.2, solving for a is a ridge regression a problem, which has a closed-form solution (given synthetic samples $\tilde{x}$ used to compute J) of $a_\theta = \mathbf{S} \cdot \sum_{i=1}^{n_s} \Phi(\tilde{x}_i) L_\theta(\tilde{x}_i)$ where $\mathbf{S} = \left(\frac{1}{n_s} \sum_{i=1}^{n_s} \Phi(\tilde{x}_i)^T \Phi(\tilde{x}_i) + \lambda I\right)^{-1}$. We then use this result in Eq. 6 to formulate the dual optimization problem as an optimization problem in θ^*:

$$\theta^* \in \arg\min_\theta \sum_{i=1}^{n_s} \underbrace{\Phi(\tilde{x}_i)^T \cdot \mathbf{S} \cdot \hat{z}_D}_{\triangleq w(\tilde{x}_i)} \cdot L_\theta(\tilde{x}_i)$$

This method, which we call *implicit*-M^{2}M, computes a weighting function $w : \Omega \to \mathbb{R}$ from the feature map Φ, private sketch $\hat{z}_D$, and regularization coefficient λ, independently of the loss. This weighting function is used to weigh the contribution of each synthetic points to the total loss. Any learning procedure, such as gradient descent, can then be applied to the re-weighted loss.

4   Experiments

We empirically evaluate the M^{2}M method on a range of data analysis tasks on artificial and real data. We perform our analyses on the LifeSci dataset, a real-world dataset of life sciences measurements ($n = 2.7 \cdot 10^4$ records and $d = 10$ attributes), which we normalize to $\Omega = [0, 1]^d$. In order to analyze the different sources of errors independently, we perform the same analyses on a uniformly sampled artificial dataset of same shape (n, d), which we call Random10. Since the training distribution ψ is equal to the empirical distribution p_X, there is no *distributional shift*, and the error observed in the results for Random10 is thus the combination of *approximation error*, *sampling error* and *the DP noise addition*.

We sketch each dataset the using RFF ($m = 200, \sigma = 1$), RACE ($R = W = 80$), and HIST (marginals of each attribute, $n_{bins} = 100$ bins of same size in $[0, 1]$), and add noise to ensure DP with privacy budget $\varepsilon \in [10^{-2}, 10^2]$, as described in Sect. 2.3. For all sketches, we split the privacy budget as $\varepsilon_{num} = 0.98\,\varepsilon$ and $\varepsilon_{den} = 0.02\,\varepsilon$. We train M^{2}M models with $n_s = 10^5$ samples, which empirically results in very low sampling error (training and testing R^2 scores almost identical). We repeat each experiment 50 times and report, for each task, the average accuracy over all runs.

An alternative to sketches is synthetic data generation (SDG), where a statistical model is fit to the real data, and so-called synthetic data are then generated by sampling from this model. We compare our results with datasets generated using three differentially private SDG methods: **DP-Copula** [24], **PrivBayes** [37], and **DP-WGAN** [36]. The latter method relies on a relaxed definition of differential privacy, $(\varepsilon, \delta)-$DP, and hence the guarantees provided are weaker. In our experiments, we use $\delta = 10^{-5}$. For each SDG and ε, we generate 10 synthetic datasets from LifeSci, and perform the tasks of interest on the synthetic data (by computing the empirical average of the functions f on the synthetic data), reporting the average over all runs.

4.1   Tasks Involving Columns in Isolation

As a first illustrative example, we consider a range of simple tasks where the function learned with M²M only concerns one attribute in isolation. For each sketch and each column in the datasets, we train a M²M model to predict (1) its mean $\frac{1}{n}\sum_{i=1}^{n} x_i$, (2) its order 2 moment $\frac{1}{n}\sum_{i=1}^{n} x_i^2$, and (3) its cumulative distribution function (CDF) in 10 equi-distant points $\left(\frac{1}{n}\sum_{i=1}^{n} I\{x_i \leq S_j\}\right)_{j=1}^{10}$. We then measure the error obtained between the predicted value and the empirical value using mean relative error (MRE) $MRE(\hat{\mu}, \mu) = \frac{|\hat{\mu}-\mu|}{\mu}$ for (1) and (2), and the Earth-Mover Distance (EMD) for (3) For each task, sketch, and dataset, we report the average error across all attributes in Fig. 3.

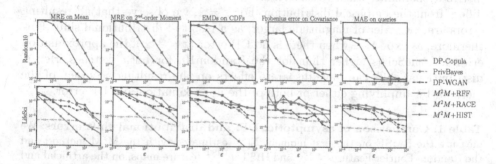

Fig. 3. Estimation of one-dimensional statistics over a random artificial dataset (top row) and LifeSci (bottom row). We estimate the mean, second-order moment, and CDF of each attribute using M²M on three sketches (RFF, RACE, and HIST), and synthetic datasets (generated using DP-Copula, PrivBayes, and DP-WGAN). We estimate the covariance matrix and the answer to a large number of random counting queries using M²M on three sketches (RFF, RACE, and HIST), and synthetic datasets (generated using DP-Copula, PrivBayes, and DP-WGAN).

We show that, in the absence of distributional shift, M²M can be used to estimate single-variable tasks with good accuracy. As expected, the HIST sketch performs well on all tasks and for both datasets, since it is specifically designed to approximate one-dimensional distributions. However, distributional shift (in LifeSci) worsens results significantly for all feature maps. This is particularly true for CDF, where the RFF and RACE feature maps result in high error, probably due to high approximation error. Comparing with synthetic data, we find that the RFF sketch compares favorably with both PrivBayes and DP-WGAN (especially when $\varepsilon \geq 1$), while the RACE sketch leads to less useful results. DP-copula datasets outperform both sketches, which is to be expected since the method explicitly estimates marginals.

We further analyze the different sources of error in Table 1. We report the mean relative error on the first moment $\mathbb{E}[X]$ obtained with either the exact sketch ($\varepsilon = +\infty$) or the private sketch with parameter $\varepsilon = 1$, for the RFF and HIST feature maps, on both datasets. For the HIST feature map and $\varepsilon = +\infty$, we find the M^2M coefficients using a small regularization $\lambda = 10^{-9}$ (for numeric stability). The error on the artificial dataset for $\varepsilon = +\infty$ is the *approximation error* of f, the irreducible error obtained when approximating f by a linear mixture of components of the feature map Φ_i. We observe that this error is low for the RFF feature map, which has strong approximation properties [32,38], and higher for the HIST sketch, which roughly approximates a function as a product of 1D piecewise constant functions. The second row ($\varepsilon = 1$) is the result of adding *DP error* to the approximation error. DP error has a negligible impact on the performances of the histogram sketch, as it is dominated by the approximation error. The opposite applies to RFF, where the DP error is 5 orders of magnitude larger. Results from the LifeSci dataset (rows 3 and 4) illustrate the impact of distributional shift, when the distribution used to generate M^2M's training set differs from the empirical distribution. For $\varepsilon = 1$, we observe that all resulting errors are one order of magnitude larger, as a result of distributional shift. Furthermore, as expected, when there is no DP error ($\varepsilon = +\infty$), the approximation error for LifeSci is higher than for the Random10, for both sketches. Hence, distributional shift can have disparate effects on the resulting accuracy of the method, by amplifying either or both of the approximation and DP error.

Table 1. Comparison of asymptotic, DP, and distributional shift errors: We measure the RMSE on the first moment $\mathbb{E}[X]$ estimated with the M^2M method and the Random Fourier Features Φ^{RFF} and HIST Φ^{HIST} feature maps, on the artificial and LifeSci datasets. We report the asymptotic error (no noise) and the error for $\varepsilon = 1$. All results are averaged over 100 trials.

Dataset	Budget	MRE for Φ^{RFF}	MRE for Φ^{HIST}
Random10	$\varepsilon = +\infty$	$6.25 \cdot 10^{-8}$	$1.87 \cdot 10^{-5}$
	$\varepsilon = 1$	$9.55 \cdot 10^{-3}$	$9.10 \cdot 10^{-4}$
LifeSci	$\varepsilon = +\infty$	$1.67 \cdot 10^{-6}$	$5.91 \cdot 10^{-5}$
	$\varepsilon = 1$	$4.20 \cdot 10^{-2}$	$3.8 \cdot 10^{-3}$

4.2 Multi-column Tasks

We evaluate M^2M on tasks that involve attributes taken together. First, we compute the covariance matrix of the dataset, $\frac{1}{n}\left((x_i - \hat{\mu}_i) \cdot (x_j - \hat{\mu}_j)\right)_{i=1,j=1}^{n,n}$, using $\hat{\mu}_i$ estimated as above. We measure the Frobenius distance between the estimated and empirical covariance matrices. Next, we perform a large number of simple counting queries $\mathrm{COUNT}(D, S)$, where the query S is defined as the conjunction of three predicates of the form $X_i \leq u$ or $X \geq l$, for three different attributes $X_i, X_{i'}, X_{i''}$. We report the Mean Absolute Error (MAE) between the real query answers and the answers predicted by M²M.

Figure 3 reports the error decrease for both tasks and on each dataset as ε increases. Similarly to the one-dimensional tasks (Fig. 3), we observe that M²M estimations perform well on the Random10 dataset, and worse on LifeSci. Except for PrivBayes, all synthetic datasets (and in particular, DP-Copula) outperform M²M. The queries use case is particularly challenging to approximate with M²M, as the target function f is not continuous. Finally, as expected, results for the HIST sketch quickly plateau for all tasks and datasets.

4.3 Logistic Regression

We use the implicit-M²M method described in Sect. 3.5 to perform logistic regression from the private sketch of a dataset. We use real-world building occupancy data [7] ($d = 6$, $n = 20,560$) with 5 continuous attributes (building characteristics) and a binary attribute (whether a building is occupied). This dataset is such that the last attribute is strongly predicted by the continuous attributes, with an AUC (area under curve) of >0.99. We normalize the continuous attributes to $[0, 1]$ and define $\Omega = [0, 1]^5 \times \{0, 1\}$ and $\psi = Unif(\Omega)$. We randomly separate the data between training (90%) and testing (10%), then sketch the training dataset using RFF ($\sigma = 1, m = 200$), RACE ($R = 80, H = 80, \sigma = 0.1$) and HIST ($n_{bins} = 20$) for a range of ε. Using implicit-M²M, we perform logistic regression on each sketch and evaluate the result on the testing dataset. We compare our results with Chaudhuri et al.'s DP-ERM [9], a dedicated method to train a logistic regression with DP using objective perturbation.

We also generate synthetic datasets using the same SDG techniques as above. We train a logistic regression using sklearn on each dataset 10 times, and measure its AUC on the test dataset. It can happen that the synthetic dataset only has one class for the last attribute; in this case we report the AUC to be 0.5.

In Fig. 4, we show that implicit-M²M compares remarkably well with DP-ERM for the RFF feature map. While it leads to higher error, the RACE feature map consistently produces an AUC of at least 0.9 for $\epsilon \geq 0.3$. Unsurprisingly, the method performs poorly on the HIST feature map ($AUC < 0.1$, not featured on the plot), which cannot, by definition, be used to estimate correlations between attributes. Importantly, models trained with implicit-M²M compare favorably with models trained on synthetic datasets using the same budget ε. As expected, the task-specific DP-ERM outperforms all other methods, but this comes at the cost of the entire budget ε. Our results suggest that implicit-M²M is a promising solution to perform sophisticated learning tasks on sketches.

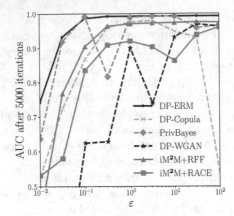

Fig. 4. AUC of a logistic regression trained from sketches on the occupancy dataset. We use implicit-M^2M to fit a logistic regression to the occupancy dataset from RFF and RACE sketches. We compare our results with the dedicated method DP-ERM and three synthetic data generation methods.

5   Future Work and Conclusion

Distributional shift occurs when the distribution used to generate M^2M's training set, ψ, differs from the data distribution p_X. This is a significant source of error in the method. We here propose a few options to reduce this error.

- Improving the approximation $\psi \approx p_X$ using the sketch. KDE sketches are built to approximate a kernel, encoding a kernel density estimate for the data distribution: $p_X(x) \approx \frac{1}{n} \sum_{i=1}^{n} \kappa(x, x_i) \approx \langle \Phi(x), \hat{z}_D \rangle$. One could thus use $\psi : x \mapsto \langle \Phi(x), \hat{z}_D \rangle$. However, the approximate distribution $\langle \Phi(x), \hat{z}_D \rangle$ can be negative and is not robust to noise addition for privacy.
- Learning a generative model on the sketch [18] that, if accurately trained, generates synthetic data similar to the sketched dataset. These synthetic records can then be used to train the M^2M model, as their distribution p_{synth} is likely to be close to p_X (or at least closer than ψ uniform). Although the synthetic records could be used directly for the learning tasks, re-accessing the data sketch through the M^2M mechanism could yield greater utility.
- Solving the loss minimization problem on the real data using a differentially private procedure. For instance, techniques such as DP-Empirical Risk Minimisation (DP-ERM) [10] could be applied – although this can be challenging, since J is non-convex. While this method is most likely the best solution to distributional shift, it requires additional privacy budget to learn the parameters of M^2M, which contradicts the idea of data summaries.

A   Proof of Theorem 1

Let J_Σ, the left-hand side of the inequality, the mean squared error between the empirical mean $\overline{f}$ and the estimation from the sketch $\widetilde{f}$. Denoting $X = (X_1, \dots, X_n)$, we have

$$J_\Sigma = \mathbb{E}_{X,\xi}\left[\left(\tfrac{1}{n}\sum_{i=1}^n f(X_i) - \langle a, \tfrac{1}{n}\left(\sum_{i=1}^n \Phi(X_i) + \xi\right)\rangle\right)^2\right]$$

$$= \mathbb{E}_{X,\xi}\left[\left(\tfrac{1}{n}\sum_{i=1}^n \left(f(X_i) - \langle a, \Phi(X_i)\rangle\right) - \tfrac{1}{n}\langle a, \xi\rangle\right)^2\right]$$

$$\overset{(i)}{=} \mathbb{E}_X\left[\left(\tfrac{1}{n}\sum_{i=1}^n \left(f(X_i) - \langle a, \Phi(X_i)\rangle\right)\right)^2\right] + \tfrac{1}{n^2}\mathbb{E}_\xi\left[\langle a, \xi\rangle^2\right]$$

$$\overset{(ii)}{=} \tfrac{n(n-1)}{n^2}\cdot\left(\mathbb{E}_X[f(X)] - \langle a, \mathbb{E}_X[\Phi(X)]\rangle\right)^2$$
$$+ \tfrac{n}{n^2}\cdot\mathbb{E}_X\left[\left(f(X) - \langle a, \Phi(X)\rangle\right)^2\right] + \|a\|_2^2\tfrac{\mathbb{V}[\xi]}{n^2}$$

where we used in (i) the independence from ξ and X and the fact that $\mathbb{E}[\xi] = 0$, and in (ii) the fact that samples $(X_i)_{1\le i\le n}$ are independent (and $\mathbb{V}[\cdot]$ denotes the variance of a random variable). Finally, we use Jensen's inequality (since $x \mapsto x^2$ is convex) to show that $(\mathbb{E}_X[f(X)] - \langle a, \mathbb{E}_X[\Phi(X)]\rangle)^2 \le \mathbb{E}_X\left[\left(f(X) - \langle a, \Phi(X)\rangle\right)^2\right]$, which concludes the proof.

B   M^{2}M Learning Procedure

Input: Target function f, private data sketch $(\hat{z}_D, n + \zeta)$ with associated
      feature map Φ and noise level σ_ξ^2, a priori distribution ψ, number of
      synthetic samples n_s, (optional) additional regularization $R = 1$.
Output: $\hat{f}$, an estimate for $\tfrac{1}{n}\sum_{i=1}^n f(x_i)$.
1 Get n_s synthetic training samples $\tilde{x}_i \sim_{i.i.d.} \psi$;
2 Set regularization parameter $\lambda = \sigma_\xi^2/(|D| + \zeta)\cdot R$;
3 Get coefficients $a = \arg\min_\alpha J(\alpha)$, using the samples $\tilde{x}_i$ as estimation for ψ;
4 **return** $\hat{f} = \langle a, \hat{z}_D\rangle \approx \tfrac{1}{n}\sum_{i=1}^n f(x_i)$.

Algorithm 1: M^{2}M learning procedure: Given a dataset sketch $\hat{z}_D$ and a target function $f : \mathbb{R}^d \to \mathbb{R}$, the procedure estimates the empirical mean of f over D.

References

1. Abowd, J.M.: The US census bureau adopts differential privacy. In: Proceedings of the 24th ACM SIGKDD International Conference on Knowledge Discovery & Data Mining, p. 2867 (2018)
2. Aktay, A., et al.: Google Covid-19 community mobility reports: anonymization process description (version 1.0). arXiv preprint arXiv:2004.04145 (2020)
3. Balog, M., Tolstikhin, I., Schölkopf, B.: Differentially private database release via kernel mean embeddings. In: International Conference on Machine Learning, pp. 414–422 (2018)
4. Barak, B., Chaudhuri, K., Dwork, C., Kale, S., McSherry, F., Talwar, K.: Privacy, accuracy, and consistency too: a holistic solution to contingency table release. In: Proceedings of the Twenty-Sixth ACM SIGMOD-SIGACT-SIGART Symposium on Principles of Database Systems, pp. 273–282 (2007)

5. Barbaro, M., Zeller, T., Hansell, S.: A face is exposed for AOL searcher no. 4417749. New York Times 9(2008), 8For (2006)
6. Blum, A., Hopcroft, J., Kannan, R.: Foundations of Data Science. Cambridge University Press, Cambridge (2020)
7. Candanedo, L.M., Feldheim, V.: Accurate occupancy detection of an office room from light, temperature, humidity and CO_2 measurements using statistical learning models. Energy Build. **112**, 28–39 (2016)
8. Chatalic, A., Schellekens, V., Houssiau, F., De Montjoye, Y.A., Jacques, L., Gribonval, R.: Compressive learning with privacy guarantees. Inf. Inference: J. IMA (iaab005) (2021). https://doi.org/10.1093/imaiai/iaab005
9. Chaudhuri, K., Monteleoni, C.: Privacy-preserving logistic regression. In: Advances in Neural Information Processing Systems, pp. 289–296 (2009)
10. Chaudhuri, K., Monteleoni, C., Sarwate, A.D.: Differentially private empirical risk minimization. J. Mach. Learn. Res. **12**(3) (2011)
11. Coleman, B., Shrivastava, A.: Sub-linear race sketches for approximate kernel density estimation on streaming data. In: Proceedings of the Web Conference 2020, pp. 1739–1749 (2020)
12. Cormode, G., Garofalakis, M., Haas, P.J., Jermaine, C.: Synopses for massive data: samples, histograms, wavelets, sketches. Found. Trends Databases **4**(1–3), 1–294 (2012)
13. Drineas, P., Kannan, R., Mahoney, M.W.: Fast Monte Carlo algorithms for matrices I: approximating matrix multiplication. SIAM J. Comput. **36**(1), 132–157 (2006)
14. Dwork, C.: Differential privacy: a survey of results. In: Agrawal, M., Du, D., Duan, Z., Li, A. (eds.) TAMC 2008. LNCS, vol. 4978, pp. 1–19. Springer, Heidelberg (2008). https://doi.org/10.1007/978-3-540-79228-4_1
15. Dwork, C., McSherry, F., Nissim, K., Smith, A.: Calibrating noise to sensitivity in private data analysis. In: Halevi, S., Rabin, T. (eds.) TCC 2006. LNCS, vol. 3876, pp. 265–284. Springer, Heidelberg (2006). https://doi.org/10.1007/11681878_14
16. Flajolet, P., Martin, G.N.: Probabilistic counting algorithms for data base applications. J. Comput. Syst. Sci. **31**(2), 182–209 (1985)
17. Gribonval, R., Blanchard, G., Keriven, N., Traonmilin, Y.: Compressive statistical learning with random feature moments. Math. Stat. Learn. **3**(2), 113–164 (2021)
18. Harder, F., Adamczewski, K., Park, M.: DP-MERF: differentially private mean embeddings with RandomFeatures for practical privacy-preserving data generation. In: Banerjee, A., Fukumizu, K. (eds.) Proceedings of the 24th International Conference on Artificial Intelligence and Statistics. Proceedings of Machine Learning Research, vol. 130, pp. 1819–1827. PMLR (2021)
19. Huang, G., Huang, G.B., Song, S., You, K.: Trends in extreme learning machines: a review. Neural Netw. **61**, 32–48 (2015). https://doi.org/10.1016/j.neunet.2014.10.001
20. Huang, G.B., Zhu, Q.Y., Siew, C.K.: Extreme learning machine: theory and applications. Neurocomputing **70**(1–3), 489–501 (2006)
21. Kenthapadi, K., Tran, T.T.: PriPeARL: a framework for privacy-preserving analytics and reporting at linkedin. In: Proceedings of the 27th ACM International Conference on Information and Knowledge Management, pp. 2183–2191 (2018)
22. Keriven, N., Bourrier, A., Gribonval, R., Pérez, P.: Sketching for large-scale learning of mixture models. Inf. Inference: J. IMA **7**(3), 447–508 (2018)
23. Keriven, N., Tremblay, N., Traonmilin, Y., Gribonval, R.: Compressive k-means. In: ICASSP (2017). https://hal.inria.fr/hal-01386077/document

24. Li, H., Xiong, L., Jiang, X.: Differentially private synthesization of multi-dimensional data using copula functions. In: Advances in Database Technology: Proceedings. International Conference on Extending Database Technology, vol. 2014, p. 475. NIH Public Access (2014)

25. Liu, F., Huang, X., Chen, Y., Suykens, J.A.: Random features for kernel approximation: a survey on algorithms, theory, and beyond. IEEE Trans. Pattern Anal. Mach. Intell. **01**, 1 (2021)

26. Misra, J., Gries, D.: Finding repeated elements. Sci. Comput. Program. **2**(2), 143–152 (1982)

27. de Montjoye, Y.A., Hidalgo, C.A., Verleysen, M., Blondel, V.D.: Unique in the crowd: the privacy bounds of human mobility. Sci. Rep. **3**, 1376 (2013)

28. de Montjoye, Y.A., Radaelli, L., Singh, V.K., et al.: Unique in the shopping mall: on the reidentifiability of credit card metadata. Science **347**(6221), 536–539 (2015)

29. Park, M., Vinaroz, M., Charusaie, M.A., Harder, F.: Polynomial magic! Hermite polynomials for private data generation. arXiv:2106.05042 [cs, stat] (2021)

30. Parzen, E.: On estimation of a probability density function and mode. Ann. Math. Stat. **33**(3), 1065–1076 (1962)

31. Qardaji, W., Yang, W., Li, N.: Priview: practical differentially private release of marginal contingency tables. In: Proceedings of the 2014 ACM SIGMOD International Conference on Management of Data, pp. 1435–1446 (2014)

32. Rahimi, A., Recht, B.: Random features for large-scale kernel machines. In: Advances in Neural Information Processing Systems, pp. 1177–1184 (2008)

33. Rahimi, A., Recht, B.: Weighted sums of random kitchen sinks: replacing minimization with randomization in learning. In: Advances in Neural Information Processing Systems, pp. 1313–1320 (2009)

34. Rudin, W.: Fourier Analysis on Groups. Interscience Publishers (1962)

35. Schellekens, V., Chatalic, A., Houssiau, F., de Montjoye, Y.A., Jacques, L., Gribonval, R.: Differentially private compressive k-means. In: 2019 IEEE International Conference on Acoustics, Speech and Signal Processing (ICASSP), ICASSP 2019, pp. 7933–7937. IEEE (2019)

36. Xie, L., Lin, K., Wang, S., Wang, F., Zhou, J.: Differentially private generative adversarial network. arXiv preprint arXiv:1802.06739 (2018)

37. Zhang, J., Cormode, G., Procopiuc, C.M., Srivastava, D., Xiao, X.: PrivBayes: private data release via Bayesian networks. ACM Trans. Database Syst. (TODS) **42**(4), 1–41 (2017)

38. Zhang, R., Lan, Y., Huang, G.B., Xu, Z.B.: Universal approximation of extreme learning machine with adaptive growth of hidden nodes. IEEE Trans. Neural Netw. Learn. Syst. **23**(2), 365–371 (2012). https://doi.org/10.1109/TNNLS.2011.2178124

JChainz: Automatic Detection
of Deserialization Vulnerabilities
for the Java Language

Luca Buccioli[1]([✉])[iD], Stefano Cristalli[3][iD], Edoardo Vignati[1][iD],
Lorenzo Nava[3][iD], Daniele Badagliacca[1][iD], Danilo Bruschi[1], Long Lu[2],
and Andrea Lanzi[1][iD]

[1] University of Milan, Milan, Italy
luca.buccioli@unimi.it
[2] Northeastern University, Boston, USA
[3] Security Pattern Inc., Milan, Italy

Abstract. In the last decade, we have seen the proliferation of code-reuse attacks that rely on deserialization of untrusted data in the context of web applications. The impact of these attacks is really important since they can be used for exposing private information of the users.

In this paper, we design a tool for automatic discovery of deserialization vulnerabilities for the Java language. Our purpose is to devise an automatic methodology that use a set of program analysis techniques and is able to output a deserialization attack chain. We test our techniques against common Java libraries used in web technology. The execution of our tool on such a dataset was able to validate the attack chains for the majority of already known vulnerabilities, and it was also able to discover multiple novel chains that represent new types of attack vectors.

1 Introduction

In the last decade, we see a propagation of code-reuse attacks in the context of web applications [4,11,14]. The impact of these attacks is important, since such vulnerabilities can be used for exposing several pieces of private information like credit card numbers, social security numbers of the common users. One example of this attack is direct to the agency Equifax, which exposes information on 143 million of US users. This attack exploits a well-known vulnerability named untrusted data deserialization in the web application context. In particular the insecure deserialization in the Apache Struts framework within a Java web application ends up in remote code execution (RCE) on Equifax web servers. The attack exploited the XML serialization data objects into textual strings and inject malicious XML payloads into Struts servers during the deserialization process. Such attacks show the need to systematically face code-reuse attack problems at research level.

More precisely to exploit this type of vulnerability, an attacker should create a custom instance of a chosen serializable class which redefines the readObject

G. Lenzini and W. Meng (Eds.): STM 2022, LNCS 13867, pp. 136–155, 2023.
https://doi.org/10.1007/978-3-031-29504-1_8

method. The object is then serialized and send to an application which will deserialize it, causing an invocation of readObject and trigger the attacker's payload. Since the attacker has complete control on the deserialized data, he can choose among all the Java classes present in the target application classpath, and manually compose them by using different techniques (e.g., wrapping instances in serialized fields, using reflection), and create an execution path that forces the deserialization process towards a specific target (e.g., execution of a dangerous method with input chosen by the attacker).

Recently researchers have published a paper that creates an automatic tool for generating a deserialization attack exploit for .NET applications [21]. Such an approach applies a practical field-sensitive taint-based dataflow analysis targeting the CIL languages. The core of such analysis leverages inter-procedural abstract interpretation based on method summaries, pointer aliasing, and efficient on-the-fly reconstruction of the control flow graph. This method is very specific for the CIL bytecode and it has not been tested on programming languages that use a different low-level representation such as Java bytecode.

Despite clear differences between Java bytecode and CIL, such representations also have similarities: both are low-level, object-oriented languages and they store objects on the heap. Even though it is tempting to create an equivalence between the two representations (e.g, applying the same analysis approach on both low-level languages), such goal is not easy to achieve since the two low-level representations work on languages with different characteristics (e.g., memory operations, safe pointer etc.). Such translation introduces issues about the exact meaning of equivalence between CIL and its translation into Java bytecode. Moreover, the translation should not introduce code artifacts that confuse the analyzer and consequently produce false positives and/or false negatives [6].

In this paper we face the problem of detection of the deserialization vulnerabilities for the Java language. Our purpose is to devise an automatic methodology that works directly on the language features and is able to output a potential deserialization attack chains. Our methodology uses a different approach compared with the one designed in [21]. In particular the analysis framework works directly on constructs of the Java language: objects/method, data type etc. and it designs an analysis which aims to discover potential attack Java deserialization chains that connect Java classes libraries. The advantage of working on the language construct is the use of semantic information that can be extracted by the rules of the language itself. As we will see in Sect. 3 such information is used to improve the precision of the analysis and reduce false positives and false negatives.

To this end we design a custom static data-flow analysis framework called JChainz, that works directly on the Java language and combines the reaching definitions and type propagation analysis for obtaining potential deserialization attack vectors, such attack vectors they need to manually validate for obtaining the final exploit. In particular our tool works in two main phases. In the first phase the system builds up a call graph and data dependency graph that contain the control-flow and data dependency information among the Java variables of the analyzed code (e.g., libraries of the target application). Goal of this phase is to select execution links between the methods of different classes (e.g. execution chains) in terms of execution call. Then, in the second phase, the system analyzes the potential attack chains and validates them. The validation is applied by propagating the variable type in the graph and marking the type inconsistencies. Such techniques can help the system to exclude the majority of false positives and select potential real attack vectors.

We test our techniques against the most common Java libraries used in web technology. More precisely we select libraries from Apache Commons Collections version 3.1 and 4.0. The Commons Collections libraries are included in a great number of projects like for example on Apache Maven Central, where we can find more than 2700 public artifacts that use such libraries. The results of our experiments show our tool was able to validate the attack chains for the majority of already known vulnerabilities, and it was also able to discover multiple novel chains [16,17], new attacks that it has been confirmed in April 2022 by yososerial research community [9].

In summary, the paper reports the following contributions:

- We present a systematic approach for discovering deserialization vulnerabilities in Java applications including the framework and libraries that is based on custom program analysis techniques.
- We design and develop a tool that is able to extract a deserialization attack vector from the Java code and help the security analyst to fix the code of the vulnerable applications. Our tool is open sourced (https://github.com/Kigorky/JChainz) for future research.
- Our experiments show the effectiveness of our approach on finding new attack vectors. We describe three new case study attacks and will also discuss the limitations of our analysis and future improvements that consider reflection and dynamic proxying techniques.

2   Background

In this section we describe background concepts for understanding the security problems of deserialization of untrusted data in Java and we provide a real attack example.

2.1   Deserialization Terminology

Java Object Serialization. Serialization is the process of encoding objects into a stream of bytes, while deserialization is the opposite operation. Java deserialization is performed by the class `Java.io.ObjectInputStream`, and in particular by its method `readObject`. A class is suitable for serialization/deserialization if the following requirements are satisfied [19]: (1) the class implements the interface `Java.io.Serializable`, (2) the class has access to the no-argument constructor of its first non-serializable superclass. A class `C` can specify custom behavior for deserialization by defining a `private void readObject` method. If present, such a method is called when an object of type C is deserialized. Other methods can be defined to control deserialization process: (1) `writeObject` is used to specify what information is written to the output stream when an object is serialized (2) `writeReplace` allows a class to nominate a replacement object to be written to the stream (3) `readResolve` allows a class to designate a replacement for the object just read from the stream.

2.2 Running Attack Example

To describe an example of a real attack, we use a real vulnerability present in Apache Common Collection libraries, and we show how an attacker can pilot a deserialization process and execute a dangerous native method. In Listing 1.2 we report the code for functions `heapify`, `siftDown` and `siftDownUsingComparator` of class `Java.util.PriorityQueue` of Java Framework. In Listings 1.3 and 1.4 we show methods `compare` of class `TransformingComparator` and method `transform` of `InvokerTransformer`, from library Apache Commons Collections 4. Listing 1.5 shows an hypothetical target class for executing a system command.

Listing 1.1. readObject in Java.util.PriorityQueue

```
private void readObject(Java.io.ObjectInputStream s) /* function a */
    throws Java.io.IOException,
           ClassNotFoundException {
    // Read in size, and any hidden stuff
    s.defaultReadObject();

    // Read in (and discard) array length
    s.readInt();
    queue = new Object[size];

    // Read in all elements.
    for (int i = 0; i < size; i++)
        queue[i] = s.readObject();

    heapify();
}
```

Listing 1.2. heapify and siftDownUsingComparator in PriorityQueue

```
private void heapify() { /* function b */
    for (int i = (size >>> 1) - 1; i >= 0; i--)
        siftDown(i, (E) queue[i]);
}

private void siftDown(int k, E x) {
    if (comparator != null)
        siftDownUsingComparator(k, x);
    else
        siftDownComparable(k, x);
}

private void siftDownUsingComparator(int k, E x) {
    int half = size >>> 1;
    while (k < half) {
        int child = (k << 1) + 1;
        Object c = queue[child];
        int right = child + 1;
        if (right < size && comparator.compare((E) c, (E) queue[right]) > 0)
            c = queue[child = right];
        if (comparator.compare(x, (E) c) <= 0)
            break;
        queue[k] = c;
        k = child;
    }
    queue[k] = x;
}
```

Listing 1.3. TransformingComparator.compare

```
public int compare(final I obj1, final I obj2) { final O value1 =
  this.transformer.transform(obj1); final O value2 =
  this.transformer.transform(obj2); return
  this.decorated.compare(value1, value2); }
```

Listing 1.4. InvokerTransformer.transform

```
public O transform(final Object input) {
  if (input == null) return null;
  try {
    final Class<?> cls = input.getClass();
    final Method method = cls.getMethod(iMethodName, iParamTypes);
    return (O) method.invoke(input, iArgs);
    ...
}
```

Listing 1.5. Command class

```
public class Command implements Serializable {
  private String command;

  public Command(String command) {
    this.command = command;
  }

  public void execute() throws IOException {
    Runtime.getRuntime().exec(command);
  }
}
```

Listing 1.6. Attack payload

```
final InvokerTransformer transformer =
  new InvokerTransformer("execute", new Class[0], new Object[0]);

final PriorityQueue<Object> queue =
  new PriorityQueue<Object>(2, new TransformingComparator(transformer));

queue.add(1);
queue.add(new Command("rm -f importantFile"));
```

Now, suppose an attacker created and serialized an object in listing 1.6. When this object is deserialized, the first method invoked after reading all the data from the priority queue is `heapify` as defined in the source code Listing 1.1 (readObject entry point of the deserialization); then `siftDownUsingComparator` is called (via `siftDown`), Listing 1.2, which uses the comparator modified by the attacker into the serialized object, in this case a `TransformingComparator`, Listing 1.6, for comparing the queue elements. The `compare` function in `TransformingComparator`, Listing 1.3, uses the field `transformer`, provided by the attacker, Listing 1.6, and calls its `transform` function on the objects being compared. At this point the `InvokerTransformer` is invoked, Listing 1.4, and such a method uses reflection to call the method with name equal to its field `iMethodName` on `input`, in this case the `Command` method. The reflection helps the attacker to invoke methods of generic classes; by crafting the deserialization input, the attacker is able to invoke method `execute` on an instance of the `Command` class with controlled parameters and execute arbitrary commands. In Listing 1.7 we report the stack trace collected at the execution of `Runtime.exec`, which contains all the Java methods invoked during the malicious deserialization event.

Listing 1.7. Stack trace of sample attack payload

```
Runtime.exec
Command.execute
Method.invoke
InvokerTransformer.transform
TransformingComparator.compare
PriorityQueue.siftDownUsingComparator
PriorityQueue.heapify
PriorityQueue.readObject
```

The attack vector described in this section is based on payload CommonsCollections2 from the ysoserial repository, used in real attacks. The only difference with the original version is the class `Command`, that we introduced for simplicity in our description. The real attack vector uses a dynamic class loading [9] as a gadget attack execution.

3   Overview

The goal of our analysis is to discover, given a specific Java library, the relationship among its classes and their methods in terms of execution. To discover such chains, as a first step we need to build a call graph that shows the relationships between methods of the analyzed classes in terms of caller and callee. Afterwards, we need to extract from the call graph, chains that reach an exit point of our interest (e.g., invoked method) and represent a potential attack vector. In Fig. 1 we depict an architectural design of our framework. More precisely, in our framework, we identify two main components: The **Finder** and the **Analyzer**. The **Finder** component starts from the Java bytecode and builds up the call graph (Scct. 3.1) of the target libraries by using the entry and exit point of any potential attack vectors (i.e., first three blocks in the diagram). Entry and exit points are defined by the deserialization process and the target attack class (Sect. 2.1). When the step is completed, the **Analyzer** component applies, for each discovered chain, the Data Dependencies Graph (Sect. 3.3) to determine the input data flow among the chain classes. In the last step of the analysis, the **Analyzer** applies a type propagation algorithm (Sect. 3.3) to exclude false positives and select the attack vectors candidates.

3.1   Call Graph Accuracy

The first challenge to solve is related to the call graph generation. A trivial solution for such a problem is to check the invoke instructions in Java bytecode, and build the call graph from them. While this represents a good starting point from our analysis, it is not sufficient to construct precise relationships among methods.

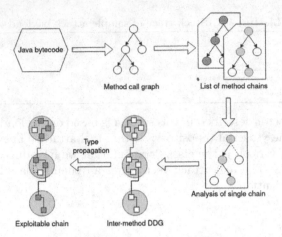

Fig. 1. Architecture of JChainz Framework

Listing 1.8. SubClass Example

```
class Example {
    SomeClass o = new SubClass();
    public a() {
        o.method();
    }
}
```

For example, consider the code in Listing 1.8. The call on o.method is performed on an instance of `SomeClass`, so the link `Example -> SomeClass` is trivial. At runtime, the instance is actually of type `SubClass`, so this link must also be considered. *Such missing information (i.e., runtime subtypes of classes and interfaces) can lead to an incomplete graph (e.g., missing chains' links), and produce false negatives since such class type is not considered and the attack vector cannot exploit its methods for executing the exploitable chain.* For this reason we need to consider such cases for building the graph, and include interface implementors and class extenders as well. More precisely, we create a link between methods in the graph only when at least one of the following conditions are satisfied:

- (1) The method's class implements the `Serializable` interface.
- (2) The callee method's class is a superclass of the caller method class.
- (3) The method has the `static` modifier.

It is important to note that all the objects (i.e., methods) that appear in the chain should be serializable. The only exceptions to such a case are invocations to methods in a non-serializable superclass (condition 2), or calls to static methods by directly invoking the method from the Java class (condition 3). For building the class call graph we use Soot [22] and we leverage Soot's capability (i.e., Soot APIs) of constructing the call graph of our input class path. In particular Soot first generates the Intermediate Representation (IR) for all the classes and their methods, and then it builds the call graph from Java invoke statements. For any invoke statement we considered class extenders and interface implementors for the analyzed callees and we label the graph according to the discovered information.

3.2   Data Type Inconsistency

Once we have built the call graph, the system extracts the execution chains and validates them. The validation process defines the input data flow from the entry point (e.g., Serializable class) of the chain till the exit point (e.g., invoke method). More precisely, the Analyzer needs to control the existence of a data flow path that depends on the input and can be used for controlling the execution of the target attack class. Such analysis needs to exclude false positives that can be raised by data type inconsistency.

Listing 1.9. Data Type Inconsistency

```
1   class Example {
2       public example() {
3           return "FOO" + "BAR";
4       }
5   }
6
7   // class StringBuilder
8   public StringBuilder append(Object var1) {
9       return this.append(String.valueOf(var1));
10  }
11
12  // class String
13  public static String valueOf(Object var0) {
14      return var0 == null ? "null" : var0.toString();
15  }
```

To see an example of data type inconsistency problem, we consider the code in Listing 1.9. In this case, we have method `Example.example`, which concatenates two strings by using the method `StringBuilder.append`, and then we have the second method `String.valueOf` that returns the string value. A correct call graph must link them, and the following chain results in a correct execution stream, as a call to `Example.example` always results in the execution of the entire chain:

C1: `Example.example -> StringBuilder.append -> String.valueOf`

Analyzing the call graph we see that `valueOf` calls the method `toString` on its `Object` parameter. In such a case to obtain a valid attack vector, we should (e.g., attacker point of view) be able to assign an instance of `Object` (i.e. anything we want) to the parameter, and proceed from there. While this reasoning would be correct if we were considering only the method `String.valueOf`, in our case such an example produce a false positive.

C2: `Example.example -> StringBuilder.append -> String.valueOf ->`
`Example.toString`

In fact such a chain misses two important pieces of information for being correctly validated: (1) the type of the parameter var0 can only be set to String (propagated from Example.example) and not to a general object, (2) moreover the string parameter is constant as defined in the class Example.example, "BAR", and consequently it cannot be modified by the attacker.

3.3 Validation Algorithm

To solve the data type inconsistency problem and validate the attack chain we need to design a custom static data-flow analysis algorithm, that combines reaching definitions and type propagation analysis and works directly on the data. The idea is to build a data dependency graph (DDG), that contains information on control flow and data dependency between variables. By propagating the variable types in the graph, we can mark type inconsistencies and remove the false positives.

Data Dependency Graph. For implementing an accurate data type propagation analysis we need to apply intra and inter method mechanisms. In the following we report how the standard algorithm works in our specific context. We first apply the intra method mechanism by considering the following steps:

- For each method in the chain, we generate the control-flow graph (CFG), and trace the data dependency starting from a reaching definitions analysis [15] (**intra-method analysis**)
- For each link in the chain M1 -> M2, corresponding to a call to M2 in the body of M1, we map the arguments in the call statement in M1 to the corresponding variables in M2 (**inter-method analysis**)

Each node in our DDG represents a particular variable defined in a specific statement. More in detail we define a node composed by a triple $(Method, Value, Unit)$, with the following parameters: (1) $Method$: represents the class and method of the current statement. (2) $Value$: represents the variable of the node. (3) $Unit$ represents the current statement. For example, considering the return statement at line 9, Listing 1.9, we can see that such statements affect two variables: this and var0. Therefore, two nodes in the DDG will be created.

We now define edges in the DDG, which represent dependency relationships between nodes. Such definitions are useful for constructing the intra-method representations. More precisely, there is an edge between node A and node B if A depends on B, the dependency is defined by the following rules:

1. A use of a variable V at a node N (with value V) depends on the definition of V at the node M
2. The definition of a variable V at a node N depends on the use of another variable U at a node M if N and M have the same unit
3. A use of a variable V at a node N (with value different U different from V) depends on the definition of U at the node M

Listing 1.10. DDG construction example

```
1   class Example {
2       public void a() {
3           String var0 = "BAR";
4           String var1 = "FOO" + var0;
5       }
6
7       public void b() {
8           new Example().c("FOOBAR");
9       }
10
11      public void c (String var2) {
12          String toPrint = var2;
13          this.a();
14          System.out.println(toPrint);
15      }
16  }
```

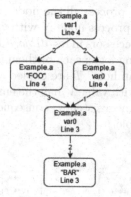

Fig. 2. intra-method DDG for Listing 1.10

In Fig. 2 we show the intra-method DDG constructed for method `Example.a` in Listing 1.10. Each edge is marked with the rule applied for data dependency. At this point, the DDG contains only information about intra-method data dependency;

After we build a intra-method we have to insert inter-method information to the graph.

To this end we follow the following strategy: when we find an invoke statement at method M_x in the chain, we check whether the callee belong to the step M_{x+1} of the chain. If this is the case, we create an inter-method edge in our DDG. In particular in our context we need to distinguish two main cases:

1. inter-method parameter call - in this case, the value of node M_x is a parameter of the method call. We track the value and make sure it is correlated with the appropriate parameter in the next method's DDG.
2. inter-method instance call - in this case, the value of node M_x is the object on which the method call is performed. Therefore, in the CFG of M_{x+1}, such object will be referenced by the `this` pointer in Java.

Type Propagation Analysis. After computing the DDG, we execute a type propagation analysis. To this end we first assign type information to each node for which the type is known, and we then propagate the information through the DDG, to detect any type inconsistencies. For this purpose, we add a dictionary to each node, named *allowed_types*. This dictionary contains an entry for each known variable at a given node in the DDG, and contains all the possible types for this variable; the types are inferred from the DDG itself.

We start with a `null` value for *allowed_types* at every node, then we initialize only nodes with no dependencies for their value, apply the following rule: for each node N with value V and no dependencies for V, we add the type of V to the *allowed_types* for V at N.

Once we statically determine the type information for each node, we navigate the graph and for each step we process all the nodes which have no dependencies with *allowed_types*. When we process node N with value V, we copy the allowed types for each variable in its successors in its *allowed_types* dictionary (duplicates are removed), with the following logic: each of them is compared with the type T of V at N; only types that "can hold" T are copied and allowed for V at N (i.e., T and supertypes). If a node with value V has the empty set as the allowed types for V, we have found a type inconsistency, consequently the data-flow through a particular node is not possible.

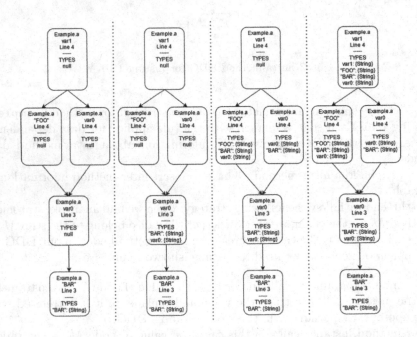

Fig. 3. Type propagation algorithm for method a in Listing 1.10. Subsequent iterations are shown from left to right

Special care is taken for inter-method links, which are handled separately. The logic of type propagation is the same, but the types are matched also on the called object and the method parameters, depending on the type of inter-method edge, described above. For instance, in the inter-method DDG shown in Fig. 3, our algorithm correctly infers type `String` for `var2`, and type `Example` for `this`. The algorithm iterates till all the nodes have been processed. Then all the nodes that are marked type inconsistency are removed from the graph. The remaining chains are marked as attack vector candidates.

4 Experimental Evaluation

In this section we present our evaluation results. We evaluate both the components of our system, the Finder and the Analyzer, measuring their effectiveness and performance. For validating the results of JChainz we also perform data analysis for discovering false positives and false negatives along with the real attack vectors. In Appendix section we also reported two case studies of new exploitation chains found by our framework. The following experiments have been run on a Debian GNU/Linux 9.11 (64 bit) machine with an Intel Xeon CPU (2.27 GHz) and 20 GB of DRAM. The code was compiled using Java OpenJDK 1.8.

4.1 Dataset

To evaluate our framework, we select the test code from the Commons Collections libraries 3.1 and 4.0 as reported in the ysoserial repository [9]. The Commons Collections packages used for the evaluation are composed of: (1) 421 classes and 3485 methods in the Commons Collections 4.0; (2) 425 classes and 3728 methods in the Commons Collections 3.1. In literature, those libraries are known to be vulnerable, thus representing an interesting target for the evaluation of our tool (e.g., Ground Truth). Due to their versatility, the Commons Collections libraries are included in a huge number of projects, for example in Apache Maven Central project. Moreover Commons Collections libraries seek to build upon the JDK classes by providing new interfaces, implementations and utilities also in a web context [7].

4.2 Finder Results

In this section we report the results of a run of the Finder on our dataset. Main aim of the Finder is to discover connections among the methods defined in a specific library. We set up the tool to perform a depth first search navigation starting from the `readObject()` custom implementations (i.e. entry points), reaching `Java.lang.reflect.Method.invoke` (i.e. exit point), which is an easy springboard for an attacker to launch arbitrary code. To help our Finder component on performing its own task we set up several parameters that control the graph's analysis exploration. In particular we define the following variables: (1) *Max depth*: maximum depth for the DFS algorithm for exploring the call graph in terms of nodes starting from a single entry point.(2) *Max chains*: maximum numbers of chains devised from a single entry point. (3) *Max seconds*: Maximum number of seconds to search for chains starting from a single entry point.

Table 1. Finder parameters

Parameter	Value
Max depth	10
Max chains	100
Max seconds	10800

In Table 1 we reported the values considered in our experiments. *The* `max` `depth` *parameter has been selected by considering the round up average depth of the already known real exploitable chains from entry points to the exit points (e.g, Common Collections in ysoserial repository).* Note that, the arguments `max_chains` and `max_seconds` are mutually exclusive parameters that interrupt the research before the full graph exploration is completed. We chose such values based on the ground truth chains parameters. In our experiments, the searching phase took approximately $40\,h$ to complete. In Table 2 we report the results of the Finder's Analysis on a single run on both libraries.

Table 2. Finder results

	CC 3.1	CC 4.0
Entry points	32	30
Active Entry Points	11	12
Total number of chains found	36	567

In particular, in the Table 2 we reported three main parameters of our results: (1) number of *Entry points* that have been statically found in the libraries. With the term "entry point", we considered the `Serializable` classes that redefine the `readObject()` method. (2) *Active Entry points*: this is a subset of entry points that contains at least one chain found by the Finder component. It is important to note that due to the time constraints of the graph exploration not all the Entry points have been analyzed in a single run of the experiment. (3) *Total number of chains found*: the number of chains for which the Finder generates a path from the entry point to an exit point (i.e., Method.invoke).

In summary, in this first run of the experiments based on the previous settings parameters, the Finder analyzed a class's call graph composed by 934300 connections (i.e., arches) for common collection 3.1 and a class's call graph composed by 519980 connections for common collection 4.0 and it was able to extract 36 chains for common collections 3.1 and 567 chains for common collections 4.0.

4.3   Analyzer Results

After the searching phase performed by the Finder component, the system sends all the found chains to the second component, the Analyzer, whose main aim is to validate them. The Analyzer starts analyzing the single chain and for each of them builds up the inter-method DDG, then it applies the data types propagation algorithm described in the previous section. *The whole analysis took approximately 27h.* In the Table 3 we report the results of these second running steps.

Table 3. Analyzer results

	CC 3.1	CC 4.0
Non-exploitable chains	22	531
Exploitable chains	8	36

As we can see from the Table 3, the Analyzer was able to exclude a large fraction of the false positive chains. In particular for the CC 4.0 the tool was able to discard 93% of the false positive chains while for the 3.1 the tool was able to discard 64% of the false positive chains. *This reduction was mainly achieved by the data type propagation analysis. In particular the average size of the Data Dependencies Graph before the pruning for common collection 3.1 was composed by 658 nodes and 1338 arches. After the pruning we obtained a graph with on average 76 nodes and 143 arches. For the common collection 4.0 we start with a graph before pruning composed by 718 nodes and 1468 arches and we obtain a graph with 138 nodes and 249 arches.*

To confirm the results we apply the manual analysis on the 44 exploitable chains found by our tool. The manual analysis reveal that 32 of the 44 exploitable chains represent false positive and 12 of the 44 chains were real attack vectors. Among the results of the exploitable chains, we first search for the ground truth exploits present in Common Collections Libraries 3.1 and 4.0. The Analyzer component was able to validate all of them with three missing exploits. Such an issue about the results depends on the fact that our tool does not handle Java Proxy classes which alter the program behavior at run-time (i.e., dynamic feature). Such dynamic gadgets are needed for exploit CC1, CC3, CC4. This represents a limitation for our tool and it will be discussed in the limitations section.

Beyond the already known exploits our tool was able to find three new real attack chains, the exploit CC7, CC8 and CC10 that have been acknowledge by the yoserial community [9] and *the other six new exploit discovered by our tool are variants of the original ones. For space limitation we report some exploits of the variants here:* https://github.com/Kigorky/JChainz/tree/main/exploits. CC7, CC8, CC10 can be considered new since they require a new exploitation technique for delivering the attack. In Appendix section we report a description on how to build up a successful attacks by using such chains.

In Table 4 we report all the known vulnerabilities validated by our tool.

Table 4. Ground Truth ChainzAnalyzer results

Vulnerability Exploit	Results
CC1 exploit	Failed
CC2 exploit	Pass
CC3 exploit	Failed
CC4 exploit	Failed
CC5 exploit	Pass
CC6 exploit	Pass

The 32 false positive chains were present in the results show the problem related to the precision of the analysis of our framework. Through the manual analysis we find out the typical false positive that our tool is affected. In particular our Analyzer is not able to process the expression of the conditions' in terms of value. In the following code we report a case of false positive found in our experiments. In particular this function is a part of the exploitable chain validated by the Analyzer.

Listing 1.11. False Positive example

```
1
2    private GeneralRange(Comparator<? super T> comparator, boolean hasLowerBound, @Nullable T
         lowerEndpoint, BoundType lowerBoundType, boolean hasUpperBound, @Nullable T upperEndpoint,
         BoundType upperBoundType) {
3        ...
4        if (hasLowerBound) {
5            comparator.compare(lowerEndpoint, lowerEndpoint);
6        }
7        ...
8    }
```

In our exploitable chain the system includes the method compare of the `Comparator` class defined as a block of the if statement (Line 5). The problem here is that the `hasLowerBound` is always set to false by the class defined in this chain and the method `comparator.compare()` will never be executed. Consequently since our chain cannot reach that method, the exploit is not feasible.

5 Limitations

Our tool is affected by some limitations, mainly due to the technical limits of the static analysis approach. The dynamic features of Java language, such as the reflection technique, are known to be an obstacle to the static code analysis. Due to the nature of these objects, the tool is not able to detect chains that could potentially be exploitable (e.g., false negatives). For example, our tool cannot handle proxy classes which alter their behavior at run-time. Some tools tried to model the static analysis over these dynamic features but this problem is still quite hard to solve [8,13].

Moreover our tool cannot handle the expression evaluation of the conditions statement (i.e., false positive). In particular such a problem could be solved by adding more precising analysis like symbolic execution. At the moment, several possibilities exist for performing symbolic execution in Java [1,2]; however, while constraint solving works well with basic types such as integers and strings, to the best of our knowledge there is currently no modeling of custom objects in OOP. If such a model were developed, then the whole search of exploitable chains could be made more accurate, by exactly solving constraints on objects and variables, and deterministically generating inputs that allow a particular chain to be executed/exploited. Another point for improving our analysis is to use a new framework for building up Java Call Graph such as [20].

6 Related Works

The most recent work related to ours is by Shcherbakov and Ballium. [21]. In their work, the authors present a tool, SerialDetector, aimed at automatic discovery of Object Injection Vulnerabilities in .NET applications and libraries. Such an approach is based on the CIL intermediate language and based its own efficacy on a practical field-sensitive taint-based dataflow analysis targeting the CLI languages. This method is very specific for the CIL bytecode and it has not been tested on programming languages that use a different low-level presentation such as Java (e.g., bytecode).

In the particular context of deserialization vulnerabilities attack, an interesting work to mention is the tool Serianalyzer by Bechler [18]. Serianalyzer uses static Java bytecode analysis to trace native method calls made during the deserialization process and it uses several heuristics to identify already known attack patterns. Although it produces many false positives, it has been used to find many of the exploits present in the ysoserial repository. In our work we decided to create a more agnostic tool that leverages the capabilities of Soot and its intermediate representation. In particular we design an automatic analysis by implementing ad-hoc data flow and type propagations analysis to discover such a tool.

On the protection side several attempts have been made for protection against attacks based on deserialization of untrusted data. Dietrich et al. [5] analyze the problem of deserialization of untrusted data not only in Java, but in several affected languages. After analyzing a few chains that cause Denial Of Service, they study in detail possible mitigation for the problem. In the specific context of Java deserialization, Cristalli et al. [3] describe a system for establishing the trusted execution path in an existing application during a learning phase, and enforcing it at run-time with analysis of stack traces in the JVM. A similar approach had been followed by Hawkins et al. [10]; their ZenIDS system uses trusted execution path validation for protection of PHP software via anomaly detection. ObjectMap [12] is a tool that aims at detecting vulnerable deserialization entry points in Java and PHP systems. Most of those works check dynamically the integrity features of the deserialization process and show a quite big run-time overhead. The final goal of our tool is to recognize and directly correct the vulnerabilities inside the Java source code.

7   Conclusion

In this paper we present a new tool, called JChainz, that is the first tool that directly work on the Java language and it is able to discover untrusted data deserialization attack vector. We present the first systematic approach for automatic creating the Deserialization attack in Java applications including the framework and libraries. We test our tool on well-known libraries and we show its effectiveness by validating results on known and new vulnerabilities. We describe three new case study attacks along with the limitations of our approach and future improvements such reflection and dynamic proxing.

Acknowledgment. This project has received funding by the Italian Ministry of Foreign Affairs and International Cooperation (grant number: PGR00814).

1   Appendix

1.1   Case Studies

By taking advantage of our tool, we discovered and exploited new chains described in the following repositories [16,17]. Each chain is composed of two main parts, the first one from the entry point to the exit-point. In this case, the exit-point is the method.invoke method. The latter exit-point allows an attacker to access and call the entire set of methods and classes available in the java classpath. The second part of the chain is composed of a gadget. In our experiments, we attached the well-known gadgets already available in the ysoserial repository, which allowed us to run arbitrary code. The gadget can be seen as an already sequence of methods for achieving a specific operation. These chains have been discovered by the Finder, filtered by the Analyzer, then manually validated and exploited.

CommonsCollections7 The payload CommonsCollections7 [17], found with the aid of our tools, consists of the following chain:

```
java.util.Hashtable.readObject
java.util.Hashtable.reconstitutionPut
collections.map.AbstractMapDecorator.equals
java.util.AbstractMap.equals
collections.map.LazyMap.get
collections.functors.ChainedTransformer.transform
collections.functors.InvokerTransformer.transform
java.lang.reflect.Method.invoke
sun.reflect.DelegatingMethodAccessorImpl.invoke
sun.reflect.NativeMethodAccessorImpl.invoke
sun.reflect.NativeMethodAccessorImpl.invoke0
java.lang.Runtime.exec
```

The chain starts in the JDK class `Hashtable`, and produces an invocation of an arbitrary system command, via `Runtime.exec`. In order to reach this result, the chain reuses the `LazyMap` gadget from chain CommonsCollections5, already part of ysoserial before our work. Therefore, the novelty of CommonsCollections7 consists of the *trigger* made of the first five methods in the chain, up to the invocation of the *gadget* with entry point `LazyMap.get`.

While the potential exploitability of the chain was confirmed by our Analyzer, we still had to build the code for the exploit. To trigger the method sequence leading to the invocation of `LazyMap.get` starting from `Hashtable`, we built an hashtable containing two instances of the `LazyMap` gadget object we wanted to reuse, with the aim of triggering comparison between the two in the hashtable upon the insertion of the second. This comparison would force the call to `equals` on the `LazyMap`, which calls method `get` and triggers the gadget.

We discovered that inserting the same object twice in the hashtable was not sufficient, as the duplicate would be recognized right away without the need of any comparison with the objects already present in the hashtable. Therefore, we fabricated two *different* instances of the `LazyMap`, but with *colliding hashes*. This was possible because it is extremely easy to obtain colliding object hashes in Java, as the hashing mechanism has not been designed for security purposes and does not make use of any cryptographic hash function. In the specific case of `LazyMap`, the hash of the entire object is calculated from the hashes of the objects in the map. Therefore, it was sufficient for us to make the keys of the `LazyMap` gadgets collide. In particular, we chose colliding String objects `"yy"` and `"zZ"`.

At this point, the `LazyMap` objects can be inserted in the `Hashtable`, which will be then serialized. When deserialized, the reconstruction of the hashtable via its custom `readObject` method will insert the two objects. The insertion of the second will trigger a comparison with the first because of the colliding hashes, starting the rest of the chain as seen above. This manual design enabled us to transform the chain found by our framework into a fully functional deserialization exploit.

CommonsCollections8. The payload CommonsCollections8 has an interesting property that differentiates it from all other previous Commons Collections payloads: its entry point (i.e. the serializable class `TreeBag`) is part of the library itself, while all other known chains have entry points in standard Java classes found in the JRE. The payload CommonsCollections8 [16] generates the following stacktrace:

```
org.apache.commons.collections4.bag.TreeBag.readObject
collections4.bag.AbstractMapBag.doReadObject
java.util.TreeMap.put
java.util.TreeMap.compare
collections4.comparators.TransformingComparator.compare
collections4.functors.InvokerTransformer.transform
java.lang.reflect.Method.invoke
```

```
sun.reflect.DelegatingMethodAccessorImpl.invoke
sun.reflect.NativeMethodAccessorImpl.invoke
sun.reflect.NativeMethodAccessorImpl.invoke0
com.sun.org.apache.xalan.(...).TemplatesImpl.newTransformer
... (TemplatesImpl gadget)
java.lang.Runtime.exec
```

This chain starts in the `TreeBag` class and leads to the execution of the `Runtime.exec` method, triggering the vulnerability in the Commons Collections 4.0 package. The contribution of this chain, like the previous one (Sect. 1.1), consists of spotting a new entry point.

The payload is composed by a `TreeBag` object built with a comparator of the type `TransformingComparator` and populated with a `TemplatesImpl` object from the ysoserial repository. During the deserialization process, the `TreeBag` class builds a new `TreeMap` object containing the attacker's comparator and passes it to the `AbstractMapBag.doReadObject` method as a parameter. At this point, the `put` method is invoked on the map object received as parameter, triggering the `compare` method call on the unsafe comparator. Starting from the `transform` method, the following operations that lead to the execution of arbitrary code are managed by the gadget from ysoserial.

References

1. Java Pathfinder. https://github.com/javapathfinder
2. Java Symbolic Execution. https://docs.angr.io/advanced-topics/java_support (2019)
3. Cristalli, S., Vignati, E., Bruschi, D., Lanzi, A.: Trusted execution path for protecting java applications against deserialization of untrusted data. In: Bailey, M., Holz, T., Stamatogiannakis, M., Ioannidis, S. (eds.) RAID 2018. LNCS, vol. 11050, pp. 445–464. Springer, Cham (2018). https://doi.org/10.1007/978-3-030-00470-5_21
4. Dahse, J., Krein, N., Holz, T.: Code reuse attacks in PHP: automated pop chain generation. In: Proceedings of the ACM Conference on Computer and Communications Security, vol. 11, pp. 42–53 (2014)
5. Dietrich, J., Jezek, K., Rasheed, S., Tahir, A., Potanin, A.: Evil pickles: dos attacks based on object-graph engineering. In: 31st European Conference on Object-Oriented Programming (ECOOP 2017). Schloss Dagstuhl-Leibniz-Zentrum fuer Informatik (2017)
6. Ferrara, P., Cortesi, A., Spoto, F.: From CIL to java bytecode: semantics-based translation for static analysis leveraging. Sci. Comput. Program. **191**, 102392 (2020)
7. The Apache Software Foundation. Java collections framework. https://commons.apache.org/proper/commons-collections/
8. Fourtounis, G., Kastrinis, G., Smaragdakis, Y.: Static analysis of java dynamic proxies. In: Proceedings of the 27th ACM SIGSOFT International Symposium on Software Testing and Analysis, ISSTA 2018, pp. 209–220, New York, NY, USA. Association for Computing Machinery (2018)
9. Frohoff, C.: ysoserial repository. https://github.com/frohoff/ysoserial (2015)

10. Hawkins, B., Demsky, B.: Zenids: introspective intrusion detection for PHP applications. In: 2017 IEEE/ACM 39th International Conference on Software Engineering (ICSE), pp. 232–243. IEEE (2017)
11. Holzinger, P., Triller, S., Bartel, A., Bodden, E.: An in-depth study of more than ten years of java exploitation, pp. 779–790 (2016)
12. Koutroumpouchos, N., Lavdanis, G., Veroni, E., Ntantogian, C., Xenakis, C.: Objectmap: detecting insecure object deserialization. In: Proceedings of the 23rd Pan-Hellenic Conference on Informatics, pp. 67–72 (2019)
13. Landman, D., Serebrenik, A., Vinju, J.J.: Challenges for static analysis of java reflection - literature review and empirical study. In: 2017 IEEE/ACM 39th International Conference on Software Engineering (ICSE), pp. 507–518 (2017)
14. Lekies, S., Kotowicz, K., Groß, S., Nava, E.V., Johns, M.: Breaking cross-site scripting mitigations via script gadgets, Code-reuse attacks for the web (2017)
15. Nielson, F., Nielson, H.R., Hankin, C.: Principles of Program Analysis. Springer Publishing Company, Incorporated, Cham (2010)
16. Authors names obfuscated. Commonscollections8 (2019). https://github.com/frohoff/ysoserial/pull/116
17. Authors names obfuscated. CommonsCollections7 (2019). https://github.com/frohoff/ysoserial/blob/master/src/main/java/ysoserial/payloads/CommonsCollections7.java
18. Bechler, M.: Serianalyzer (2017). https://github.com/mbechler/serianalyzer
19. Oracle Corporation. The serializable interface (2017). https://docs.oracle.com/javase/8/docs/platform/serialization/spec/serial-arch.html#a4539
20. Santos, J.C., Jones, R.A., Ashiogwu, C., Mirakhorli, M.: Serialization-aware call graph construction. In: Proceedings of the 10th ACM SIGPLAN International Workshop on the State Of the Art in Program Analysis, SOAP 2021, pp. 37–42. Association for Computing Machinery, New York (2021)
21. Shcherbakov, M., Balliu, M.: Serialdetector: principled and practical exploration of object injection vulnerabilities for the web. In: Network and Distributed Systems Security (NDSS) Symposium 202121–24 February 2021 (2021)
22. Vallée-Rai, R., Co, P., Gagnon, E., Hendren, L., Lam, P., Sundaresan, V.: Soot - a java bytecode optimization framework. In: Proceedings of the 1999 Conference of the Centre for Advanced Studies on Collaborative Research, CASCON '99, p. 13. IBM Press (1999)

FlowADGAN: Adversarial Learning for Deep Anomaly Network Intrusion Detection

Pan Wang[1]([✉]), Zeyi Li[1], Xiaokang Zhou[2,3], Chunhua Su[4],
and Weizheng Wang[5]

[1] School of Modern Posts, Nanjing University of Posts and Telecommunications,
Nanjing, China
wangpan@njupt.edu.cn
[2] Faculty of Data Science, Shiga University, Hikone 522-8522, Japan
[3] RIKEN Center for Advanced Intelligence Project, RIKEN, Tokyo 103-0027, Japan
[4] Division of Computer Science, University of Aizu, Fukushima 965–8580, Japan
[5] Department of Computer Science, City University of Hong Kong,
Hong Kong SAR, China

Abstract. Due to the increasingly evolved attacks on the Internet, especially IoT, 5G, and vehicle networking, a robust Network Intrusion Detection System (NIDS) has gained increasing attention from academic and industrial communities. Anomaly-based intrusion detection algorithms aim to detect unexpected deviations in the expected network behaviour, thus detecting unknown or novel attacks compared to signature-based methods. Deep Anomaly Detection (DAD) technologies have attracted much attention for their ability to detect unknown attacks without manually building the traffic behaviours profile. However, low recall rates and high dependencies on data labels still hinder the development of DAD technologies. Inspired by the successes of Generative Adversarial Networks (GANs) for detecting anomalies in the area of Computer Vision and Images, we have proposed a deep end-to-end architecture called FlowADGAN for detecting anomalies in NIDS. Unlike traditional GAN-based NIDS methods that usually construct Generator (G) and Discriminator (D) based on vanilla GAN, the proposed architecture is composed of a flow encoder-decoder-encoder for G, and a flow encoder for D. FlowADGAN can learn a latent flow feature space of G so that the latent space better captures the normality underlying the network traffic data. We conduct several experimental comparisons with existing machine learning algorithms like One-Class SVM, LOF, and PCA and existing deep learning methods, including AutoEncoder and VAE, on three public datasets, NSL-KDD CICIDS2017 and UNSW-NB15. The evaluation results show that FlowADGAN can significantly improve the performance of the anomaly-based NIDS.

Keywords: Anomaly detection · Unsupervised learning · Generative adversarial network · Intrusion detection system

G. Lenzini and W. Meng (Eds.): STM 2022, LNCS 13867, pp. 156–174, 2023.
https://doi.org/10.1007/978-3-031-29504-1_9

1   Introduction

With the rapid advancement in the Internet, especially IoT, 5G and vehicle networking, cyber security has gained increasing attention of both the academic and industrial community, which has motivated many researchers to design and develop effective Network Intrusion Detection systems (NIDS). Generally, intrusion detection technologies fall into two genres: *misuse-based* detection (also known as signature-based detection) and *anomaly-based* detection (also known as behavior-based detection). Misuse-based detection systems can identify known attacks based on the predefined signatures effectively and efficiently. These approaches usually score high detection capabilities and low false-positive rates when facing known attacks [10]. However, frequent manual signature maintenance by security experts is required. However, even more importantly, they cannot effectively adapt the evolving attacks like *zero-day* attacks, which are novel attacks that cannot be matched to the known patterns [26]. Correspondingly, anomaly-based IDSs can capture any deviation from characteristics of normal network behavior to mitigate the problem above. These deviations are called *anomalies* [8]. Therefore, they are more suitable for detecting unknown or novel attacks than misuse-based approaches. It is worth mentioning that anomaly-based detection algorithms highly depend on the ability to characterize expected and, consequently, anomalous behaviors. Hence, anomaly-based algorithms usually suffer from high false alarms than misuse-based methods.

Many Machine Learning (ML) algorithms have been employed for developing Anomaly Detection (AD) models, such as *Clustering, Neighbor-based, Density-based, Statistical, Angle-based, and Classification-based* [2,33]. Nonetheless, there are many challenges for ML-based AD algorithms. First, network traffic data is complex, high dimensional, and non-linear. Second, normal network behaviour evolves rapidly. Third, the notion of anomaly is subjective and depends on the application domain and context. And then, anomaly labels are rare, usually done by human experts manually. Finally, the boundary between normal and anomalies is often not precise [8]. Fortunately, as a revolutionary, innovative paradigm, Deep Learning (DL) has gained great success in computer vision, image and speech. DL leverages automatic feature learning to achieve better performance, avoiding task-specific engineering and lots of prior knowledge, and thus, become popular now.

Recently, Deep Anomaly Detection (DAD) technologies have attracted much attention for their ability to detect unknown attacks without manually building the traffic behaviors profile [21]. This technique automatically learns hierarchical discriminative features from historical traffic data of massive normal traffic and minimal anomalous traffic, reducing network traffic complexity and discovering implicit correlations between data without human intervention. In addition, DAD is more robust in detecting zero-day attacks and adapting to evolving systems. Nonetheless, DAD technologies still are more or less limited by the following challenges [22]: (1) low anomaly detection recall due to the lack of well-designed DL models and threshold selection algorithms, (2) high dependence on anomalous traffic data labels, (3)detecting traffic behavior deviations is challenging, as the boundary between normal and anomalous traffic is often not clear and precise. The last one is the interpretability of DL models to make the AD algorithms trustworthy.

As a promising paradigm, Generative Adversarial Network (GAN) has been applied in video generation, image captioning and text translation with its excellent automatic features extraction by adversarial learning. Subsequently, the GAN-based anomaly detection technique emerges quickly as one popular DAD approach, which generally aims to learn a latent feature space of a generative network G so that the latent space well captures the normality underlying the given data. The anomaly score is defined by the residual between the actual instance and the generated one [29].

This paper proposed a GAN-based AD algorithm named FlowADGAN using adversarial learning to learn the underlying distribution of the network flow data. Structurally, the model is constructed using a convolutional encoder and a deconvolutional decoder. Then we select the three most representative datasets and validate the superiority of the designed model compared to the other five methods under the three datasets. The last and the main contribution of this paper is:

1. FlowADGAN uses an internal structure of an adversarial production network consisting of three convolutional encoders and a deconvolution decoder. The hidden vector computed by two encoders can represent the normal network flow features well, with better results than the traditional GAN methods.
2. This algorithm proposes an anomaly detection algorithm based on improved threshold selection. The algorithm can reasonably obtain the normal and abnormal flow boundary using the kernel function and set it as the threshold.
3. FlowADGAN exploits a flow-encoder to transform the original feature space into a more condensed and semantically rich embedding space. In this space, the model structure implements the function of reconstructing the aggregated features of the network, which applies to many different feature datasets and has good scalability.
4. Three network traffic dataset comparison experiments evaluate the proposed method, achieving good results.

2 Related Works

2.1 Review of Anomaly Detection Algorithms

The first wave of *Anomaly Detection* Algorithms was on the basis of ML algorithms, which can be divided into six families: [11]: Clustering [30], Neighbor-based [31], Density-based [5], Statistical [14], Angle-based [13], and Classification-based [4,16]. However, the first wave posed several challenges: (1) low recall rate, (2) high-dimensional or not-independent data, (3) data efficient learning of normality/abnormality; (4) noise resillient.

The second wave is based on DL algorithms since 2015 due to the superior performance applied in the area of Computer Vison/Image/Speech. We can simply summarize three apparent trends: (1) more novel and advanced DL algorithms are adopted, such as MLP/CNN/AE/RNN even GAN/GNN etc. (2) learning methods adopted by DL models are gradually from supervised to semi-supervised, weak-supervised and unsupervised learning. (3) the anomaly score learning approaches are moving from separately learning to end-to-end including more precise threshold selection algorithms.

Table 1. A summary of works on ML/DL based AD method

Reference	Family Classifier	learning style	Model Input	Datasets	Anomaly Score
Li et al. [15]	OCSVM	unsupervised	flow features	1999 DARPA	classification
Camacho et al. [7]	PCA	semi-supervised	flow features	private	RE
Paulauskas et al. [23]	LOF	unsupervised	flow features	private	density deviation
AutoIDS [12]	cascading AE	semi-supervised	flow features	NSL-KDD/private	RE
Kitsune [19]	ensemble AE	unsupervised	flow features	private	RE
Pratomo et al. [24]	AE	semi-supervised	byte frequency	UNSW-NB15	z-score
Zavrak et al. [36]	AE/VAE	semi-supervised	flow features	CICIDS2017	RE/RP(Reconstruct Prob)
CANnolo [17]	LSTM-AE	semi-supervised	byte sequence	Alfa Romeo	RE/Mahalanobis distance
MENSA [32]	AE-GAN	semi-supervised	flow features	MTU	adversarial loss
Akcay et al. [3]	DL(cov-GAN)	semi-supervised	Images	CIFAR10;MINST	discriminative loss

2.2 Deep Anomaly Detection Algorithms in IDS

Deep Anomaly Detection (DAD) techniques for short, aims at learning hierarchical discriminative features or anomaly score via deep neural networks for the sake of anomaly detection. In summary, there are two main genres applied in DAD-based IDS: AE-based and GAN-based, which will be introduced in detail as following subsections. We summarize some related works on ML/DL based AD methods listed in Table 1.

Autoencoder-Based AD Methods in IDS. As a popular deep structure for AD, AE and its genres have played a very important role during this decades. M.Gharib *et al.* [12] presented *AutoIDS*, a network anomaly detector by cascading a sparse AE and AE to increase accuracy and decrease the time complexity, in which anomalous flows are distinguished from normal ones by the first detector and the second one is only used for difficult samples that the first detector is not confident about. Y.Mirsky *et al.* [18] proposed a plug and play NIDS called *Kitsune* which utilised an ensemble of AEs to collectively detect anomalous traffic on the local network in an unsupervised online manner. The evaluations showed that *Kitsune* can detect various attacks with a performance comparable to offline anomaly detectors, even on a Raspberry PI. S.Zavrak *et al.* [35] applied AE and VAE to detect anomalous network traffic from flow-based data. The experimental results show that VAE outperforms AE and One-Class SVM on CICIDS2017 dataset. B.Abolhasanzadeh *et al.* [1] proposed an AE based IDS approach from the point of view of dimensionality reduction to detect anomalous network behavior.

In summary, The AE family has several advantages: the methods are simple, easy to design and general for different data types.

GAN-Based AD Methods in IDS. GAN-based anomaly detection methods are generally used to learn the latent feature space by adversarial learning of G (Generative Network) and D (Discriminative Network) so that the latent space can capture the normality of the given data well. Schlegl [29] proposed AnoGAN, a deep convolutional GAN to learn multiple normal anatomical variants, accompanied by a new anomaly scoring scheme based on a mapping from image space to latent space. Zenati [37] utilizes a recently developed GAN model for anomaly detection and achieves state-of-the-art performance on image and NIDS datasets. Subsequently, Schlegl [28] proposed fast AnoGAN (f-AnoGAN), a GAN-based semi-supervised learning method that can identify anomalous images and fragments. Akcay [3] introduced a new anomaly detection model called *GANomaly* that uses conditional GAN (CGAN) to jointly learn the generation of high-dimensional image space and inference of the latent space with given conditions.

However, the algorithm is only applicable to images as the input of the model, which can not fit the feature information of the traffic well. In short, GAN models [9] have shown excellent ability in generating actual instances, especially on image/video data. However, the currently GAN-based AD methods still have the following limitations. Firstly, most existing GAN-based AD models are only suitable for images/videos, not for network traffic. Secondly, existing GAN-based AD algorithms are not capable of extracting and retaining network information well when mapping high-dimensional network traffic features to low-dimensional features, which resulting in a weak ability to represent features in the mapping of normal network data to low dimensional latent features, hence, leading to a poorly discriminative ability for network anomalies.

3   Methodology

3.1   FlowADGAN Model Design

Figure 1 illustrates an overview of our method and architecture. Figure 2 shows the model parameters and output in the encoder and decoder.

The generator learns the normal network flow representation and reconstructs the input network flow through the Encoder and Decoder networks.

Consider that a data set D_{train} contains K normal flow characteristic records, $D_{train} = \{x_1, \ldots, x_k\}$. We use normal flow characteristic records to train the GAN network.

At the same time, we construct a test data set $D_{test} = \{(x_1, y_1), \ldots, (x_j, y_j)\}$, where y_i belongs to the label of network traffic, $y_i \in 0, 1$. D_{test} contains M normal test samples and N malicious test samples, $(M \gg N)$.

Our goal is to model and learn internal characteristics from D_{train}, and then identify abnormal Data in D_{test}. The FlowADGAN algorithm learns the distribution of normal traffic and minimizes the abnormal fraction S(x). When a given test sample is encountered, its S(x) is calculated. If S(x)> the threshold, it is considered an abnormal flow.

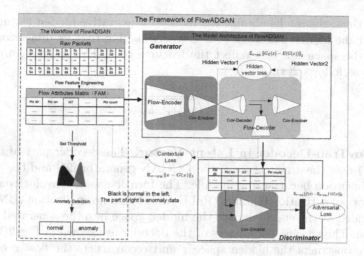

Fig. 1. Workflow of our method and model architecture.

Learning a Fast Mapping from Flow Features to Encodings in the Subspace. Flow-encoder is very important in the input part and is the main contribution of this paper; it can be applied to a variety of different network traffic data. Furthermore, the model will uniformly map the multi-dimensional feature space to the 32-dimensional feature space (v) when inputting network flow features to obtain aggregated features. The model uses a learnable linear layer as the feature mapping function ϕ. Weight matrix parameters $W \in \mathbb{R}^{d \times D}$ are used to obtain a new embedding stream feature space with powerful representation capabilities. A simple linear transformation can also ensure the scalability of the model. The data object x is transferred to a new embedding space through the feature mapping function $\phi(x)$, as follows:

$$\phi(x) = \begin{bmatrix} \mathbf{W}(1,1)x(1) + \mathbf{W}(1,2)x(2) + \cdots + \mathbf{W}(1,D)x(D) \\ \mathbf{W}(2,1)x(1) + \mathbf{W}(2,2)x(2) + \cdots + \mathbf{W}(2,D)x(D) \\ \cdots \\ \mathbf{W}(d,1)x(1) + \mathbf{W}(d,2)x(2) + \cdots + \mathbf{W}(d,D)x(D) \end{bmatrix} \tag{1}$$

where $w(i,j)$ represents the element in the $i-th$ row and $j-th$ column of the matrix w, and $x(i)$ represents the $i-th$ element of the vector x. Each dimension can be regarded as a linear pattern (combination) of the original feature space. The mapping function $\phi(x)$ will effectively aggregate the network flow features into a new embedded flow space v and then pass it to the generator coding network (GE) for spatial mapping.

Using Cov-Encoder in Subspace. The flow embedding space newly generated in GE uses a convolutional layer and performs Batch Normalization (BN) and function activation LeakyReLU() respectively. By compressing the embedded flow space v into a hidden space z, the dimensionality of v is reduced. At the same time, the model obtains the most representative hidden features. z is also called the hidden feature of x, $Z \in \mathbb{R}^d$. Assuming that its dimension is the smallest, the space can best reflect the features of x. The specific convolution process is as follows 2(a):

$$Z = \frac{x - \hat{E}[x]}{\sqrt{\hat{\mathrm{Var}}[x] + \epsilon}} \tag{2}$$

Using CovTran-Decoder in Latent Vector. The decoder part of the generator (G) uses the architecture of the DCGAN generator [25] and the network feature dimension backtracking module. The former uses a convolutional transposed layer, an activation function (ReLU) and batch normalization (BN). Tanh layer is added at the end to decode the hidden space into the generated embedding space. The specific parameters of the decoder are shown in Fig. 2(b). This method reconstructs the hidden space z and reconstructs the flow v as $\hat{v}$. On this basis, the dimensionality is expanded using linear changes through the mapping function $\psi(x)$ to become a new space $\hat{x}$. The whole process is called spatial backtracking, where $\hat{v} = \mathrm{GE}(z)$.

$$\psi(x) = \begin{bmatrix} \mathbf{W}(1,1)v(1) + \mathbf{W}(1,2)v(2) + \cdots + \mathbf{W}(1,D)v(D) \\ \mathbf{W}(2,1)v(1) + \mathbf{W}(2,2)v(2) + \cdots + \mathbf{W}(2,D)v(D) \\ \cdots \\ \mathbf{W}(d,1)v(1) + \mathbf{W}(d,2)v(2) + \cdots + \mathbf{W}(d,D)v(D) \end{bmatrix} \tag{3}$$

Using Cov-Encoder in Reconstructed Network Flows. In this part, the model compresses the flow data space $\hat{v}$ reconstructed by the neural network. $\hat{v}$ reconstruct the flow space compression to find its characteristics, specifically expressed: $\hat{z} = \mathrm{E}(\hat{v})$. The dimension of the hidden space vector $\hat{z}$ is the same as the dimension of z so that the distance can be calculated later. This subnetwork is special in the proposed method, where it can represent the hidden features in the reconstruction space. Unlike previous methods, [36] based on VAE, the distance between the latent space and the original stream data space is minimized by hiding features. The sub-network GE minimizes the distance through parameterized explicit learning.

Discrimination Between Generated Flows and Real Flows. The goal is to classify input x and output as true or false respectively. This sub-network is a standard discriminator network introduced in DCGAN.

3.2   FlowADGAN Pipeline

From Fig. 1, the formal principle of this sub-network is as follows: The generator G first reads the input network flow feature data x, where $x \in R \mid$ (w× featural of network flow) and forwards it to a layer of dimensional compression fully connected v, and then passes it to the code Network GE. Using a convolutional layer, and then performing batch norm and activation function, respectively, the dimension of v is reduced by compressing it into a vector z. z is also called the hidden feature of network flow. These characteristics best represent normal network flow. The decoder part of the generator network G uses a ConvTranspose layer, activation functions $ReLU$, and batch quota together with a $tanh$ layer at the end. This method scales the vector z and reconstructs the flow v as $\hat{v}$. Finally, the potential network embedded traffic space is restored to a network space with the exact dimensions as the input.

The second sub-network is the Encoder, which compresses the network flow data $\hat{v}$ reconstructed by the neural network. GE is compressed downward to $\hat{v}$, and its characteristic representation $\hat{z} = E(\hat{v})$ is found. The dimension of the vector $\hat{z}$ is the same as the dimension of z so that the distance can be calculated later. The third sub-network is the discriminator network D, whose goal is to classify input x and output $\hat{x}$ as true or false, respectively.

3.3   Algorithm

We assume that when anomaly flow passes through the generator. The generator cannot reconstruct the abnormal flow, and This is because the network is trained on normal samples, and its parameterized modelling is not suitable for generating abnormal samples. The reconstruction failure $\hat{v}$ means that the encoder network $GE(\hat{v})$ cannot be mapped to a vector $\hat{z}$ typically, resulting in an abnormal distance between z and $\hat{z}$. See details for Algorithm 1 .

Algorithm 1: Unsupervised anomaly detection algorithm based on FlowADGAN

---

input : Normal real data, $X_\mathcal{T} = \{x_0, x_1, x_2, \ldots\ldots x_n\}$ is the stream feature vector of the first network data stream, i is the number of iterations, N is the total number of training sets. m is the size of mini batch.

1   *Use the generative network to calculate fake flow, v and $\hat{v}$;*
2   **for** $i \leftarrow 1$ **to** N **do**
3   Enter the real network traffic into the discriminator to determine the $(pred - real)$;
4   this $\rightarrow$FindCompress$(z = \phi(x))$;
5   this $\rightarrow$FindCompress$(z = GE(v))$;
6   this $\rightarrow$FindCompress$(\hat{v} = GD(z))$;
7   this $\rightarrow$FindCompress$(\hat{x} = \psi(\hat{v}))$;
8   this $\rightarrow$FindCompress$(\hat{z} = GE(\hat{v}))$;
9   Send the fake flow generated by the generator to the discriminator to get the $(pred - fake)$;
10     $Loss_{adv} = \mathbb{E}_{x \sim p_X} \|f(x) - \mathbb{E}_{x \sim p_X} f(G(x)\|_2$;
11     $Loss_{con} = \mathbb{E}_{x \sim p_X} \|x - G(x)\|_1$;
12     $Loss_{hiddenloss} = \mathbb{E}_{x \sim p_X} \|G_E(x) - E(G(x))\|_2$;
13     $Loss = w_{adv}Loss_{adv} + w_{con}Loss_{con} + w_{hidden}Loss_{hidden}$;
14   Update parameters ;
15   Start the backward by BCELoss in the discriminator and get errord ;
16   Update parameters ;

The loss function of the discriminator is defined as follows:

$$Loss_{adv} = \mathbb{E}_{x \sim p_X} \|f(x) - \mathbb{E}_{x \sim p_X} f(G(x)\|_2 \qquad (4)$$

However, since there is only adversarial loss, the generator is not optimized to learn contextual information about the input data. The loss functions of normal flow and generated flow are defined as follows:

$$Loss_{con} = \mathbb{E}_{x \sim p_\mathbf{x}} \|x - G(x)\|_1 \tag{5}$$

In this paper, an additional loss is added to constrain the two hidden vectors. The loss function is used to minimize the distance between two hidden vectors. The definition of this function is as follows:

$$Loss_{hidden} = \mathbb{E}_{x \sim p_\mathbf{x}} \|G_E(x) - E(G(x))\|_2 \tag{6}$$

We construct the overall function by adjusting the weight parameters. Details as follows:

$$Loss = w_{adv}Loss_{adv} + w_{con}Loss_{con} + w_{hidden}Loss_{hidden} \tag{7}$$

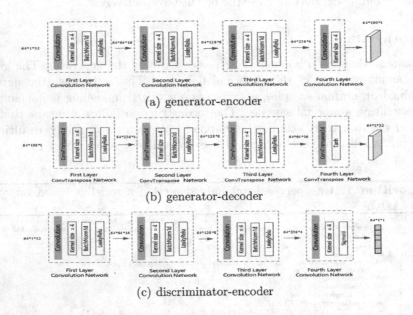

(a) generator-encoder

(b) generator-decoder

(c) discriminator-encoder

Fig. 2. The parameter of FlowADGAN.

3.4 Anomaly Scores

The anomaly score calculation method in this paper uses formula 1 to calculate the anomaly of network traffic for scoring. Therefore, for a test sample x, anomaly score A(x) is defined as:

$$S(x) = \|G_E(\phi(\mathbf{x})) - G_E(\hat{v})\|_1 \tag{8}$$

In order to choose a suitable threshold, we need to normalize the anomaly scores of a single test sample x in the test set, thus generating a set anomaly S. Threrefore, we apply S(x) mapping to anomaly scores at [0, 1].

$$S_i = \frac{S_i - S_{\min}(x)}{S_{\max}(x) - S_{\min}(x)} \tag{9}$$

3.5   Threshold Selection

The threshold selection method is as follows. The first is the choice of validation set data. The experiment selects a certain proportion of normal and malicious traffic in an orderly manner. Then, the validation set is fed into the anomaly detection model. Calculate the reconstruction distance of each normal flow sample and malicious flow sample in the validation set according to the formulas 9 and 8. Finally, the probability density and kernel function 13 are obtained according to the reconstruction distance of the verification set, and the threshold is determined.

In validation set, there are n abnormal scores as follows: $\{S_1, \quad S_2, \ldots, S_n\}$. Assume that the cumulative distribution function of the sample data is F(x).

$$F(x_{i-1} < x < x_i) = \int_{x_{i-1}}^{x_i} f(x)dx \tag{10}$$

And the probability density function is f(x):

$$f(x_i) = \lim_{h \to 0} \frac{F(x_i + h) - F(x_i - h)}{2h} \tag{11}$$

Introduce the empirical distribution function of the cumulative distribution function and substitute this function into f(x):

$$f(x_i) = \lim_{h \to 0} \frac{1}{2nh} \sum_{i=1}^{n} 1_{x_i - h \le x_j \le x_i + h} \tag{12}$$

After determining h, f(x) can be transformed into:

$$f(x) = \frac{1}{2nh} \sum_{i=1}^{n} 1_{x-h \le x_i \le x+h} = \frac{1}{2nh} \sum_{i=1}^{n} K\left(\frac{|x - x_i|}{h}\right) \tag{13}$$

Through the kernel density function, we calculate the two functions of normal traffic and malicious traffic respectively, and get the intersection. Turn the intersection into a threshold.

4   Evaluation

In this section, we evaluate FlowADGAN in terms of its detection and run-time performance. We will describe the data sets and experimental setup and then finally demonstrate our evaluation results.

4.1   Evaluation Settings and Chosen Datasets

The data sets we select need to meet the experimental requirement. As the three classic NIDS data sets, NSL-KDD [27], CICIDS2017, and UNSW-NB15 [20] are desirable. In addition, as the benchmark data sets, their experimental evaluation results are of great significance.

The experimental environment is AMD Ryzen 3600, 16GB RAM, NVIDIA GTX 1660, CUDA 7.5, CDNN10.5. In this paper, Python3 is the primary programming language.

The following is a description of the evaluation index system: precision, Recall, F1, Accuracy and AUC as evaluation metrics.

Marc avg means that each type of sample is given equal weight. For example, in this article, the Macro Average accuracy index p is defined as:

$$P = \frac{P_{normal} + P_{malware}}{2} \tag{14}$$

Weighted avg is to use the proportion of the sample size of each category in the total number of samples in all categories as the weight. For example, in this article, the Weighted Average accuracy index p is defined as:

$$P = \frac{N_{normal}}{N_{normal} + N_{malware}} * P_{normal} + \frac{N_{malware}}{N_{normal} + N_{malware}} * P_{malware} \tag{15}$$

If time-related metrics (detection time/exection time) are not considered, it can be explained that these are all dependent on hardware resources.

4.2   Ablation Study

We first do an ablation study using the UNSW-NB15 dataset in this subsection.

Table 2 shows that if the model aggregates the flow features first, the representative elements will be represented as a matrix in the subspace. Then the spatial mapping will be performed. From F1, the results may not be suitable if only the module used for image anomaly detection is directly migrated to detect network traffic. The weighted average of F1 for the model with the aggregated feature module is 0.4334 higher, and the AUC is increased by 0.0646. Therefore, the module designed in this paper can significantly improve network flow feature detection performance.

Table 2. Module comparison of improved models

Table 3. GAN compared in three Datasets

Model	Progressive	Original	Aggregation of Features
Precision	weighted avg	0.7647	0.9877
	marc avg	0.6890	0.5072
Recall	weighted avg	0.3852	0.5004
	marc avg	0.5063	0.6827
F1	weighted avg	0.2236	0.6570
	marc avg	0.2881	0.3482
AUC		0.7808	0.8454

Model(AUC)	NB15	CICIDS	NSL-KDD
FlowADGAN	0.8454	0.7461	0.9810
f-anogan	0.7215	0.6929	0.9572
EGBAD	0.5638	0.7365	0.9372

The experiment compares the model designed with recent models based on GAN to detect anomaly flows in a network environment in a comprehensive manner. Since models need to be evaluated globally, AUC has become a leading model evaluation indicator. FlowADGAN performs better than the other two models in the Table 3. On the UNSW-NB15 data set, the AUC value of this model reaches 0.8454, which is 0.1239 higher than f-anogan. Our model proposed is more effective than EGBAD. The CICIDS2017 data set has a vast amount of data. Therefore, with sufficient training, although the AUC value of FlowADGAN is the highest, the difference between the three types of gan models is not significant. Also, on the NSL-KDD data set, the model proposed in this paper has a higher AUC than the other two models. From an overall point of view, this model works best. Therefore, we can conclude that the FlowADGAN model clusters the features before inputting the features, which can refine the feature form of the data and, at the same time, use the loss function to optimize and constrain the hidden features. In the next stage of comparison, we use FlowADGAN as the representative of the GAN series model to compare with other family's algorithm models.

4.3 Performance Evaluation

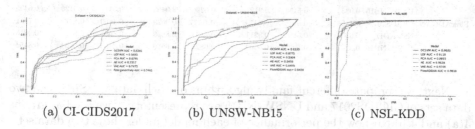

(a) CI-CIDS2017   (b) UNSW-NB15   (c) NSL-KDD

Fig. 3. ROC of Anomaly Detection Algorithms

This experiment selects representative algorithm data sets for testing in different algorithm families. In Table 3(c), as AUC is the overall indicator of the evaluation model, we can see that the algorithm proposed in this article has a better effect on each data set than other algorithms. As shown in Fig. 3(b), in the NSL-KDD data set, although FlowADGAN shows the best effect, it is not much different from other algorithms. Since the NSL-KDD data set was constructed earlier, the flow of this data set does not have the timeliness of the current network. At the same time, the network flow under this data set is relatively simple. Therefore, each model can learn its characteristics and achieve better results. Our algorithm shows stable performance on the UNSW-NB15 and CICIDS2017 data sets. From Fig. 3(a), its AUC value of 0.8454 is second only to the first LOF. This algorithm can better learn data distribution under the condition of three constraint functions and a small training sample size simultaneously. The CICIDS2017 data set is the latest data set of the three data sets, and it is also the most extensive data set. This data set fully shows the characteristics of different models. The deep learning model also exerts its advantages. From Fig. 3(a), the AUC value of the deep learning model exceeds 0.73, while other machine learning algorithms are between 0.5 and 0.6. Thus, the FlowADGAN proposed in this article is still in a leading position in deep learning algorithms.

Table 4. Evaluation results.

Model	Dataset	Precision		Recall		F1		Accuracy	AUC
		weighted avg	marc avg	weighted avg	marc avg	weighted avg	marc avg		
OCSVM	CICIDS2017	0.9831	0.5031	0.5545	0.5773	0.7044	0.3686	0.5545	0.6341
	NSL-KDD	0.9899	0.5433	0.9059	0.9050	0.9423	0.5550	0.9059	0.9680
	UNSW-NB15	0.9831	0.5058	0.7730	0.6039	0.8631	0.4538	0.7730	0.8339
PCA	CICIDS2017	0.9806	0.5081	0.9866	0.5030	0.9836	0.5037	0.9866	0.6795
	NSL-KDD	0.9901	0.5422	0.9015	0.9106	0.9398	0.5519	0.9015	0.9603
	UNSW-NB15	0.9831	0.5067	0.8014	0.6077	0.8808	0.4643	0.8014	0.5904
LOF	CICIDS2017	0.9814	0.5418	0.9882	0.5100	0.9846	0.5149	0.9882	**0.5483**
	NSL-KDD	0.9893	0.5336	0.8816	0.8768	0.9285	0.5319	0.8816	0.9118
	UNSW-NB15	0.9879	0.5242	0.8573	0.8113	0.9144	0.5092	0.8573	**0.8775**
AE	CICIDS2017	0.9848	0.5132	0.8385	0.6823	0.9032	0.4859	0.8385	0.7357
	NSL-KDD	0.9902	0.5154	0.6952	0.8392	0.8106	0.4392	0.6952	0.9628
	UNSW-NB15	0.9753	0.4968	0.1969	0.4495	0.3181	0.1691	0.1969	**0.5456**
VAE	CICIDS2017	0.9841	0.5420	0.9614	0.6447	0.9720	0.5610	0.9614	0.7475
	NSL-KDD	0.9903	0.5511	0.9192	0.9236	0.9499	0.5714	0.9192	0.9706
	UNSW-NB15	0.9832	0.5154	0.9125	0.6206	0.9453	0.5111	0.9125	0.6455
FlowADGAN	CICIDS2017	0.9850	0.5135	0.8379	**0.6877**	0.9029	0.4863	0.8379	0.7461
	NSL-KDD	0.9904	0.5256	0.8211	0.8998	0.8928	0.4993	0.8211	0.9810
	UNSW-NB15	0.9877	0.5072	0.5004	0.6827	0.6570	0.3482	0.5004	0.8454

Next, we compare different family algorithms on each data set. Since the two data sets of CICIDS2017 and UNSW-NB15 are representative, we use bar graphs 4(a) and 4(b) to show the performance of each model under these two data sets. In the field of anomaly detection, the model will have the problem of judging benign samples as malicious. However, we believe that the problem of the model judging malicious samples as benign samples is more serious. Therefore, the number of false flows determined as proper flows should be as small as possible. So we need to consider that the benign accuracy rate and the negative recall

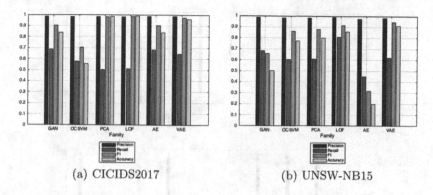

(a) CICIDS2017   (b) UNSW-NB15

Fig. 4. Metric scores of Anomaly Detection Algorithms

rate have great weight. This experiment shows the weighted average precision rate and the macro average recall rate.

FlowADGANs' accuracy rate exceeds 98% mainly because the ratio of normal flow to malicious flow in the training set is 100:1. Therefore, most of the normal flow can be well-identified. Their accuracy rates are high. On the other hand, the recall rate is lower in each data set. The reason is that too much malicious flow is identified as normal flow. Therefore, whether anomaly detection can detect anomalous flow, the recall index is essential. It can be seen from Fig. 4(a) that on the CICIDS2017 data set, as a representative of deep learning algorithms, the FlowADGAN recall rate is higher than other algorithms. Therefore, the model has a relatively strong ability to detect abnormal flow.

In addition, on the UNSW-NB15 data set, the density-based outlier algorithm LOF performed extremely abnormally. The LOF algorithm compares the density of the sample points around two sample points [6]. Therefore, LOF has more advantages in learning small-scale data samples. However, the algorithm will be more unstable when converting data sets or increasing the size. See from the Fig. 4(b), on the CICIDS2017 data set, the AUC value of LOF is only 0.5483, indicating that the model is not practical. Therefore, the effect of encountering unknown flow may not be good.

4.4   Experiment Discussion

In this experiment, we select two newer data sets, and box plots show the abnormal scores of each type of malicious flow. Figure 5 demonstrates that the model does not have a good recognition effect on Exploits, Fuzzers, and Reconnaissance. Fuzzers are an attack that attempts to suspend programs or networks by providing randomly generated data. Reconnaissance is an attack that includes all attacks that can simulate information gathering. Both types of attacks are generated by simulation, which is inherently random. Therefore, the characteristics of these two types of flow are similar to normal flow, and it is difficult for the model to distinguish them. Exploits are vulnerability attacks. The attacker

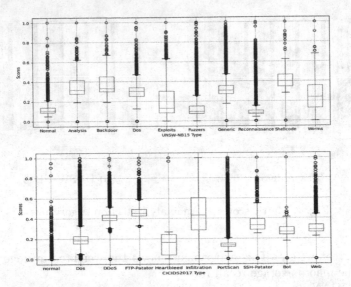

Fig. 5. Box-line figure of malicious flow classification.

knows the security issues in the operating system or software and exploits this knowledge by exploiting the vulnerability. The flow of these vulnerabilities is generated along with the flow of software applications. Therefore, the flow characteristics of these vulnerabilities are similar to the typical flow characteristics in the software operation process, which makes the model difficult to identify. Attacks like Heartbleed are loopholes in the ssh protocol, and there are too few samples available for testing. Therefore, the recognition effect of FlowADGAN is not good. On the data set UNSW-NB15, the model in this article has too little normal sample training, so the abnormal score of normal flow will be slightly higher than the CICIDS2017 data set. Brute-force cracking of this type of attack is often through repeated trial and error by enumerating exhaustive methods. Therefore, there will be apparent manifestations in features such as flow duration that allow the model to distinguish the difference from the normal flow easily.

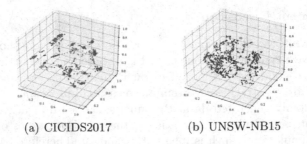

(a) CICIDS2017   (b) UNSW-NB15

Fig. 6. Visualization of hidden vectors.

Based on the Fig. 6, we select a ratio of 100:1 between normal and malicious flow. The FlowADGAN model can distinguish the flow well. In the last part of the experiment, we also performed model interpretability work on the UNSW-NB15 dataset. We use the ATON [34] method to explain the outlier results of anomaly detection. The ATON algorithm is used for post-mortem interpretation. It puts the abnormal data that has been obtained into the ATON algorithm, thereby explaining the contribution of each feature under the modified model and data set. The figure below shows that features such as sttl and ct-srv-dst play a more important role in the model.

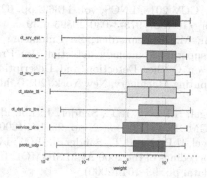

Fig. 7. Interpretation of the model and important features in UNSW-NB15

5 Conclusion

In this paper, we propose a FlowADGAN anomaly detection algorithm to solve the problem of complex network structure and the inability to identify new malicious attacks. The model creatively proposes a practical aggregation module that can be applied to data features of different dimensions as input. At the same time, the module can aggregate the autocorrelation features and then carry out the convolution coding, which can retain the abnormal feature information to a greater extent. The algorithm directly learns the spatial features after aggregation, completes the spatial backtracking, and forms multiple constraints, which enhances the stability of the model and facilitates training. Many experiments show that FlowADGAN performs better than other traditional algorithms on public intrusion detection data sets NSL-KDD, UNSW-NB15 and CICIDS2017.

However, research on unsupervised flow detection algorithms is rare and immature. Although unsupervised algorithms have made some achievements in this paper, there is still much room for improvement. In the future, we still have much work to do to improve the intrusion detection model.

Acknowledgment. This work was supported by the National Science Foundation of China (61972211) and National Key R&D Program of China.

References

1. Abolhasanzadeh, B.: Nonlinear dimensionality reduction for intrusion detection using auto-encoder bottleneck features. In: 2015 7th Conference on Information and Knowledge Technology (IKT), pp. 15 (2015). https://doi.org/10.1109/IKT.2015.7288799
2. Ahmed, M., Naser Mahmood, A., Hu, J.: A survey of network anomaly detection techniques. J. Netw. Comput. Appl. 60(C), 19–31 (2016). https://doi.org/10.1016/j.jnca.2015.11.016
3. Akcay, S., Atapour-Abarghouei, A., Breckon, T.P.: GANomaly: semi-supervised anomaly detection via adversarial training. In: Jawahar, C.V., Li, H., Mori, G., Schindler, K. (eds.) ACCV 2018. LNCS, vol. 11363, pp. 622–637. Springer, Cham (2019). https://doi.org/10.1007/978-3-030-20893-6_39
4. Amer, M., Goldstein, M., Abdennadher, S.: Enhancing one-class support vector machines for unsupervised anomaly detection. In: Proceedings of the ACM SIGKDD Workshop on Outlier Detection and Description, ODD'13, pp. 8–15. Association for Computing Machinery, New York (2013). https://doi.org/10.1145/2500853.2500857
5. Breunig, M.M., Kriegel, H.P., Ng, R.T., Sander, J.: Lof: identifying density-based local outliers. In: SIGMOD, vol. 29, no, 2, pp. 93–104 (2000)
6. Breunig, M.M., Kriegel, H.P., Ng, R.T., Sander, J.: LOF: identifying density-based local outliers. In: Proceedings of the 2000 ACM SIGMOD International Conference on Management of data, pp. 93–104 (2000)
7. Camacho, J., Pérez-Villegas, A., García-Teodoro, P., Maciá-Fernández, G.: PCA-based multivariate statistical network monitoring for anomaly detection. Comput. Secur. 59, 118–137 (2016)
8. Chandola, V., Banerjee, A., Kumar, V.: Anomaly detection: a survey. ACM Comput. Surv. 41(3), 1–58 (2009). https://doi.org/10.1145/1541880.1541882
9. Creswell, A., White, T., Dumoulin, V., Arulkumaran, K., Sengupta, B., Bharath, A.A.: Generative adversarial networks: an overview. IEEE Signal Process. Mag. 35(1), 53–65 (2018)
10. Depren, O., Topallar, M., Anarim, E., Ciliz, M.K.: An intelligent intrusion detection system (ids) for anomaly and misuse detection in computer networks. Expert Syst. Appl. 29(4), 713–722 (2005). https://doi.org/10.1016/j.eswa.2005.05.002, https://www.sciencedirect.com/science/article/pii/S0957417405000989
11. Falcão, F., et al.: Quantitative comparison of unsupervised anomaly detection algorithms for intrusion detection. In: Proceedings of the 34th ACM/SIGAPP Symposium on Applied Computing, SAC'19, pp. 318–327. Association for Computing Machinery, New York (2019). https://doi.org/10.1145/3297280.3297314
12. Gharib, M., Mohammadi, B., Dastgerdi, S.H., Sabokrou, M.: Autoids: auto-encoder based method for intrusion detection system. ArXiv abs/1911.03306 (2019)
13. Kriegel, H.P., Schubert, M., Zimek, A.: Angle-based outlier detection in high-dimensional data. In: Proceedings of the 14th ACM SIGKDD International Conference on Knowledge Discovery and Data Mining, KDD'08, pp. 444–452. Association for Computing Machinery, New York (2008). https://doi.org/10.1145/1401890.1401946
14. Kwitt, R., Hofmann, U.: Unsupervised anomaly detection in network traffic by means of robust PCA. In: 2007 International Multi-Conference on Computing in the Global Information Technology (ICCGI'07), p. 37 (2007). https://doi.org/10.1109/ICCGI.2007.62

15. Li, K.L., Huang, H.K., Tian, S.F., Xu, W.: Improving one-class SVM for anomaly detection. In: Proceedings of the 2003 International Conference on Machine Learning and Cybernetics (IEEE Cat. No. 03EX693), vol. 5, pp. 3077–3081. IEEE (2003)

16. Liu, F.T., Ting, K.M., Zhou, Z.H.: Isolation forest. In: 2008 8th IEEE International Conference on Data Mining, pp. 413–422 (2008). https://doi.org/10.1109/ICDM.2008.17

17. Longari, S., Nova Valcarcel, D.H., Zago, M., Carminati, M., Zanero, S.: Cannolo: An anomaly detection system based on LSTM autoencoders for controller area network. IEEE Trans. Netw. Serv. Manage. 18(2), 1913–1924 (2021). https://doi.org/10.1109/TNSM.2020.3038991

18. Mirsky, Y., Doitshman, T., Elovici, Y., Shabtai, A.: Kitsune: an ensemble of autoencoders for online network intrusion detection. ArXiv abs/1802.09089 (2018)

19. Mirsky, Y., Doitshman, T., Elovici, Y., Shabtai, A.: Kitsune: an ensemble of autoencoders for online network intrusion detection. arXiv preprint. arXiv:1802.09089 (2018)

20. Moustafa, N., Slay, J.: UNSW-NB15: a comprehensive data set for network intrusion detection systems (UNSW-NB15 network data set). In: 2015 Military Communications and Information Systems Conference (MilCIS), pp. 1–6. IEEE (2015)

21. Pang, G., Shen, C., Cao, L., Hengel, A.V.D.: Deep learning for anomaly detection: a review. ACM Comput. Surv. 54(2), 1–38 (2021). https://doi.org/10.1145/3439950

22. Pang, G., Shen, C., Cao, L., Hengel, A.V.D.: Deep learning for anomaly detection: a review. ACM Comput. Surv. (CSUR) 54(2), 1–38 (2021)

23. Paulauskas, N., Bagdonas, A.F.: Local outlier factor use for the network flow anomaly detection. Secur. Commun. Netw. 8(18), 4203–4212 (2015)

24. Pratomo, B.A., Burnap, P., Theodorakopoulos, G.: Unsupervised approach for detecting low rate attacks on network traffic with autoencoder. In: 2018 International Conference on Cyber Security and Protection of Digital Services (Cyber Security), pp. 1–8 (2018). https://doi.org/10.1109/CyberSecPODS.2018.8560678

25. Radford, A., Metz, L., Chintala, S.: Unsupervised representation learning with deep convolutional generative adversarial networks. arXiv preprint. arXiv:1511.06434 (2015)

26. Ramaswamy, S., Rastogi, R., Shim, K.: Efficient algorithms for mining outliers from large data sets. In: SIGMOD, vol. 29, no. 2, pp. 427–438 (2000). https://doi.org/10.1145/335191.335437

27. Revathi, S., Malathi, A.: A detailed analysis on NSL-KDD dataset using various machine learning techniques for intrusion detection. Int. J. Eng. Res. Technol. (IJERT) 2(12), 1848–1853 (2013)

28. Schlegl, T., Seeböck, P., Waldstein, S.M., Langs, G., Schmidt-Erfurth, U.: f-AnoGAN: Fast unsupervised anomaly detection with generative adversarial networks. Med. Image Anal. 54, 30–44 (2019)

29. Schlegl, T., Seeböck, P., Waldstein, S.M., Schmidt-Erfurth, U., Langs, G.: Unsupervised anomaly detection with generative adversarial networks to guide marker discovery. In: Niethammer, M., et al. (eds.) IPMI 2017. LNCS, vol. 10265, pp. 146–157. Springer, Cham (2017). https://doi.org/10.1007/978-3-319-59050-9_12

30. Schubert, E., Koos, A., Emrich, T., Züfle, A., Schmid, K.A., Zimek, A.: A framework for clustering uncertain data. In: Proceedings of the VLDB Endowment, vol. 8, no. 12, pp. 1976–1979 (2015). https://doi.org/10.14778/2824032.2824115

31. Shubair, A., Ramadass, S., Altyeb, A.A.: kENFIS: kNN-based evolving neuro-fuzzy inference system for computer worms detection. J. Intell. Fuzzy Syst. 26(4), 1893–1908 (2014)

32. Siniosoglou, I., Radoglou-Grammatikis, P., Efstathopoulos, G., Fouliras, P., Sarigiannidis, P.: A unified deep learning anomaly detection and classification approach for smart grid environments. IEEE Trans. Netw. Serv. Manage. **18**(2), 1137–1151 (2021). https://doi.org/10.1109/TNSM.2021.3078381
33. Usama, M., et al.: Unsupervised machine learning for networking: techniques, applications and research challenges. IEEE Access **7**, 65579–65615 (2019). https://doi.org/10.1109/ACCESS.2019.2916648
34. Xu, H., et al.: Beyond outlier detection: interpreting outliers by attention-guided triplet deviation network. In: Proceedings of the Web Conference 2021 (WWW'21). ACM (2021)
35. Zavrak, S., İskefiyeli, M.: Anomaly-based intrusion detection from network flow features using variational autoencoder. IEEE Access **8**, 108346–108358 (2020). https://doi.org/10.1109/ACCESS.2020.3001350
36. Zavrak, S., Iskefiyeli, M.: Anomaly-based intrusion detection from network flow features using variational autoencoder. IEEE Access **8**, 108346–108358 (2020)
37. Zenati, H., Foo, C.S., Lecouat, B., Manek, G., Chandrasekhar, V.R.: Efficient GAN-based anomaly detection. arXiv preprint. arXiv:1802.06222 (2018)

Trust and Security

The Relevance of Consent in the Digital Age: A Consideration of Its Origins and Its Fit for Digital Application

Marietjie Botes[1,2(✉)] [iD]

[1] SnT Interdisciplinary Centre for Security, Reliability and Trust, Luxembourg University, Esch-sur-Alzette, Luxembourg
marietjiebotes1@gmail.com
[2] School of Law, University of Kwa-Zulu Natal, Durban, South Africa

Abstract. Consent originated in the 1800s to protect incarcerated prisoners against unwanted medical treatment and was later formalized in the Nuremberg Code in response to harmful medical experiments that was conducted on prisoners of war during World War II. These co-called ethical principles was later reinforced and extended to protect the control and decisional power that individuals need over their bodies in The Belmont Report. Today these ethical consent principles are codified in laws such as the GDPR. Considering that these ethical consent principles was developed around biomedical treatments and experiments, it begs the question whether these same principles are still relevant and can be successfully applied in a digital environment. This paper critically considers the application of the original ethical consent principles in the digital age and highlights certain critical challenges. The aim of the paper is to draw attention to the fact that the concept of consent and whether it can still be applied ethically in a digital environment must be considered first before digital consent models or consent automation tools are developed, because such a consideration will have a critical impact on how these tools must be developed to remain, not only legal, but also ethical and subsequently sustainable.

Keywords: Legal consent · Ethical consent · Digital consent · Origins of biomedical consent · Relevance of consent

1 The Origins of Consent

1.1 From Biomedical Treatment to Clinical Research

The first legally binding obligation to obtain informed consent for medical treatment was contained in a directive issued by the Prussian Minister of Interior Affairs in 1892 which directed that tuberculosis treatment may not be administered against the will of

G. Lenzini and W. Meng (Eds.): STM 2022, LNCS 13867, pp. 177–188, 2023.
https://doi.org/10.1007/978-3-031-29504-1_10

the patient (in this specific case incarcerated patients) [1]. After the Neisser-case[1] in 1900 a further directive prohibited the participation of minors or non-competent people from non-therapeutic clinical research and demanded that clinicians obtain "unambiguous consent" after possible negative consequences of the study have been disclosed to potential participants [1].

The first internationally recognized informed consent guideline as it is known and still applied today is contained in The Nuremberg Code that was handed down by the court after the prosecution of Nazi medical doctors for their involvement in controversial and harmful research activities conducted on concentration camp prisoners during World War II [2]. This code emphasized the voluntary nature of consent and stated that "any element of force, fraud, deceit, duress, overreaching," "constraint" or "coercion" are strictly forbidden in the context of consent [2]. This code further defined consent by stating that the "individual who initiates, directs or engages in the experiment" must also provide participants with "sufficient knowledge and comprehension about the elements" of the study such as the nature, duration, purpose, method, and meaning of the experiment, including information about any expected inconveniences, hazards, and effects – and that this duty cannot be delegated to anybody else [2].

But these information requirements were soon disregarded by doctors who conducted longitudinal Syphilis studies on patients who were completely unaware of the fact that they were part of a study that lasted from 1932–1972 [3]. To further establish informed consent practices and to protect an individual's ability to exercise control and decisional power over his or her body the Belmont Report was issued as an outcry against the paternalistic practices that dominated doctor–patient relationships in the 1900s, including ongoing abuses of people participating in biomedical research with the goal to enable individuals to exercise control and decisional power over their bodies [4].

The main aim of the development of these consent instruments was to protect people who found themselves in a situation where they may suffer possible harm, if they continue to engage in a certain relationship with another, because of not having sufficient information to help him or her to decide for themselves whether they want to continue in these circumstances. This information deficiency caused a power imbalance that opens possibilities for exploitation or harm. Today power and informational imbalances persist, but are we now urged to consider the relevance of consent in view of the nature of digital technologies and how these technologies can be used to give control and decisional power back to internet users.

---

[1] Albert Neisser (1855–1916) was a Prussian professor of dermatology and venerology at the University of Breslau, who conducted clinical studies on the treatment of syphilis. For the development of an effective vaccination he injected patients (mostly sexual workers), admitted with other medical conditions, with a cell-free serum. The vaccination, after a public outcry on approx. 600 cases collected by a psychiatrist Albert Moll, was deemed as 'unsuccessful.' In 1898 the Royal Disciplinary Court condemned the activities of Neisser. The main argument was not the questionable scientific background of the studies but the lack of consent from the patients.

1.2 Consent as Legal Right

Informed consent requirements are now codified in law. The General Data Protection Regulation (GDPR) states specifically that natural persons "should have control of their own personal data" [5]. However, the GDPR took a human rights approach to privacy, leaving the technical and administrative aspects of how to implement the envisioned consent requirements behind. In this context, it strives to provide the individual with greater control and autonomy over his or her decision-making power when it entails their personal information, whilst placing more emphasis on the obligations of the data controller to operate in a transparent and lawful manner [6]. In addition, the GDPR requires from the data controller to provide the data subject with a bunch of information as listed in Articles 13 and 14 of the GDPR to enable the data subject to adequately assess the circumstances and consequences of the data processing activity they are about to give their consent to [7]. Up to this point the goals and ambitions of the GDPR seems to be aligned with the ethical consent principles of the Nuremberg Code and Belmont Report. However, it also seems, as will be discussed below, that too much emphasize is increasingly being placed on the legal requirements of consent, as opposed to the ethical principles of consent and what consent is supposed to achieve in society. One of the differentiating factors between ethical and legal consent is the issue of comprehension, and this is where ethical consent principles supplement the legal requirements of consent as stated in the GDPR. The Nuremberg Code, including ethical guidelines following thereafter, such as the Belmont Report, Declaration of Helsinki, and guidelines issued by the Council for International Organizations and Medical Sciences (CIOMS), specifically provide those participants should not only be provided with "sufficient knowledge" or information, but that participants' "comprehension about the elements" of the study must also be ensured [2]. The CIOMS guidelines, for example, places specific emphasize on the capacity of the patient to consideration information with reference to his or her age, maturity, and cognitive ability, because these factors may directly impact a patient's ability to *understand* information, which consequently influences whether his consent can be considered to be ethical based on his comprehension of the information [31]. Although the GDPR makes reference to the concept of understanding, as part of the principle of transparency of an individual's right to information, this concept is only described to the extent that information must be "concise, easily accessible, and *easy to understand*" [32]. However, this principle does not seem to impose on the data controller the same obligation as the above ethical guidelines imposes on the biomedical researcher to ensure that the data subject *fully comprehend* the information given to the participant prior to providing consent, failing which such consent will be deemed unethical. For consent to be considered legally valid in terms of the GDPR, consent must be "freely given, specific, informed, and unambiguous" with no mention of the ethical obligation to ensure comprehension of the information by the data subject [33]. In essence legal consent only requires the provision of the listed information to an individual and that consent be expressed by such individual, whilst ethical consent requires the extra condition of ensuring that the participant truly *understand* information before consenting.

1.3 Ethical v Legal Consent

The authoritative authors on consent ethics, Beauchamp and Childress, distinguish between three elements of informed consent: 1) consent threshold elements that include competence and voluntariness; 2) information elements such as disclosure, provision of recommendations, and understanding; and 3) consent elements such as decision and authorization [8]. From this it is clear that ethical consent entails a much richer engagement with the data subject than legal consent. Ethical consent requires that an individual truly understand or comprehend information that is given to him or her, as opposed to merely being provided with information (albeit in an easy-to-understand format), as legally required, and that an individual is afforded sufficient time to consider such information to enable the individual to make a truly considered, informed, and subsequent autonomous decision.

Unfortunately, consent processes in the digital world frequently fail to live up to these underlying ethical values that justified its creation in the first place. In case of medical treatment patients have to sign a consent form before surgery can start and after they were made to understand the risks and consequences involved in such surgery, which may allow a patient to rather forego the surgery if the patient deems the risks too high. Patients may seek a second, or even a third opinion from various medical practitioners specializing in the same field as the one in which the patient requires treatment before making a final decision. In these circumstances is may be reasonably acceptable, based on standardized specialist training which medical specialists undergo, to expect that certain medical specialists should be able to provide certain standard services related to their specialist field, such as thoracic surgery. A second or third opinion may thus relate to a practitioner's success record with a specific treatment regime, whether the patient finds personal rapport with a certain practitioner, or whether the patient trust the practitioner more than others and is thus more accepting of his or her advice. But in the digital world a user often has to consent to certain privacy agreements to access specific digital platforms or services in a fairly take-it-or-leave-it approach, which platforms increasingly collect and use users' personal information for covert purposes, making consent unethical, illegal, and futile [9]. Although it may be argued that users, like patients, also have the option to forego using a specific application, or rather consider using other applications or websites, it is often the case that users want to use a specific application, such as Twitter of Facebook, because of their network, scale, functions, and the fact that most of the user's contacts are also using this specific application. In these circumstances the user is either forced to consent to whatever terms and conditions the specific platform offers, or face being excluded from accessing it, while different social media applications may not offer the same network of possible contacts, scale, or functions.

Digital consent processes usually only disclose the potential uses the company has for an individual's personal data, as opposed to empower the individual to consider his or her options and control the use of their personal information. While this may be legally valid consent, it fails to live up to consent's ethical goals. Considering the origins of consent, as discussed above, consent is supposed to transform the relationship between an individual in an informationally vulnerable position, and someone in an informationally powerful position, into a more balanced relationship with regards to power by improving information equality. In case of digital consent processes, a gap opens up between legally valid consent and this ethically or morally transformative consent. Because of the purely techno-legal approach to online consenting, digital consent has been repeatedly criticized as being an "unworkable, empty procedural act" [9]. Everyone is so busy automating and digitizing existing consent concepts and models that nobody questions whether the original concept of consent for medical treatment and research is still relevant, or whether it is still applicable in the digital world. For us to take full advantage of emerging technologies, we need to consciously integrate ethics and moral philosophy, in addition to other relevant disciplines, into the implementation of digital consenting by critically reconsidering some foundational ethical and moral concepts. As a moral concept consent plays a morally transformative role in interpersonal interactions as discussed above. In the medical context, the informational gap that exist between the doctor or clinical researcher and patient or participant, which gap could be narrowed by the provision of information to the patient or participant to enable them to understand the risks and consequences involved to the extent that they feel satisfied enough to make an informed choice. However, in the digital world it is questionable how this knowledge gap can be bridged with the provision of more information to the online user, considering that in this context even the informational gaps between computer scientists and software engineers grows larger by the day, including the informational gaps between the creators of algorithms and their AI creations. In an effort to comply with the transparency requirements that is not only foundational to ethical consent, but also required in terms of the GDPR, digital information exchange is based on notifications when information of users is electronically collected and used, which notification also provide the user with a choice of whether to accept or reject the collection and use as suggested by this digital information exchange [10]. However, due to the unpredictability or unimaginable use of data by AI systems, these technologies exacerbate existing legal-ethical issues with online consent [11].

Almost based on the same principle of transformative consent and the rebalancing of informational inequalities, Miller and Wertheimer also refers to the cooperative nature of consent in which there are "fair and predictable standards underlying the consent transaction for both the consenter and consentee" [12]. They emphasize the bilateral nature of the consent transaction as if consent is a commercial transaction between two contracting parties. In this context if may be the case that the concept of consent has evolved to more resemble commercial transactions in which contracting parties must mutually agree on the terms and conditions of the transaction, including their respective performances in terms of the agreement for the agreement and subsequent transaction to be legal. Consent as "digital transaction" considered in this context may differ from consent that was given by individuals for medical treatment or research in which the patient or participant agreed to be subjected to certain activities, such as surgery or experimentation, without them having to perform something in return to the doctor or researcher. In contrast, so-called digital consent transactions are increasingly concluded between online platforms or service providers and users which entail not only users' permission to be subjected to certain activities without any counter performance, but consent in these cases are the legal basis for the collection of users' personal or behavioral information. Accordingly users do counter perform like a contract party as opposed to only permitting to be subjected to activities performed on them by a third party. Considering the interests of both online parties, the legal obligation of transparency that is placed on the data processor in terms of the GDPR, as well as the ethical goals to ensure the rebalance of power through information exchange becomes critical when the consenter and consentee look more and more like contracting parties with legally enforceable rights and obligations that flows from the terms and conditions contained in the agreement between them. The difference between consent and a contract always needs explanation, but unfortunately falls outside the scope of this article, save to say that this evolution of consent increases the demands for transparent information that is fully comprehended by the consenter.

Related to the idea that consent is a transaction that requires the cooperation and effective communication of the rights and obligations between the consenter and consentee is the issue that it is common to perpetuate on-line consent and that the consent transaction may subsequently not be a once-off transaction. Due to the nature of users' continuous interaction with on-line platforms, such as their social media use, these digital transactions will have to be ongoing over an extended period of time to effectively account for the new information that is collected from on-line users each time they access and use these platforms. This is remarkably different from obtaining consent in a medical treatment context where the patient gives consent for a specific and singular treatment plan or medical action. A single consent or permission for this type of "transaction" is thus sufficient, as opposed to the perpetuating nature of the digital consent transaction.

2 Digital Consent

Robust ethical digital consent seems realistically speaking almost impossible to achieve. In addition, many of the privacy debates are heavily influenced by socio-technical, economic and even political cultures. For example, Yao-Huai found that contemporary notions of privacy in China constitute a combination of traditional Chinese emphases on the importance of the family as well as the state in which scenario privacy is justified as an instrumental good, and a more Western orientated individual rights such as the right to privacy where privacy is considered to be an intrinsic good [13]. Warf also found China, Burma/Myanmar, Vietnam and Iran to be the countries in the world with the most severe internet censorship across the globe, with countries such as Russia, Belarus, Pakistan, and the Arab world following shortly behind them [14]. Governmental influenced concepts of privacy and internet censorship directly and negatively affects people's right to autonomously choose their interactions or so-called digital consent transactions with and on the internet. The process of standardizing digital consent is similarly influenced by social and political dimensions. To prevent the prioritization of one cultural view above another any standardization proposal must acknowledge and accommodate differences in cultural dimensions of informational privacy and consenting [15]. In this regard the current one size fits all consent mechanisms found in the digital world does not appreciate these cultural differences and how it affects consenting interfaces.

Solove [16] identified three main challenges to the management of digital information using a consent model, which challenges we shall expand on below from an ethical perspective.

2.1 Habituation and Information Overload

It is impossible for users to read all the privacy notices and their terms and conditions contained in the privacy policies they encounter on-line on a daily basis. Bravo-Lillo et al. call this phenomenon "pop-up fatigue" or "habituation" to describe the tendency of users to increasingly ignore relevant information in circumstances where they are confronted with the same or similar conditions, such as cookie banners, repeatedly over an extended period of time [17]. This habituative behavior towards online consent forms is resulting in users only clicking through privacy and consent notices instead of attentively reading and understanding the terms and condition to inform their consent choices which calls into question whether the technical consent resulting from this process is truly ethical [18]. Surprisingly very few differences have been found between the use of online consent documents and paper based and personally distributed consent documents. Varnhagen et al. suggested that researchers should not focus on the modality of the consent documents, but rather on ways to encourage attentive engagement with consent documents – whether on-line or not [19]. Even when information was presented in text format Obar and Oeldorf-Hirsch have shown that users often provide their consent after only having skim-read or blatantly ignoring the information entirely, simply so that they could access their desired service as soon as possible [25]. Considering that both on-line as well as traditional paper-based consent forms lead to some degree of habituation, it may be the case that the digital environment merely exaggerated the problem of habituation that was initially triggered and given momentum by information overload [20]. Schermer, Custers, and van der Hof, also argues that information overload, especially in the absence of meaningful choices (see Sect. 2.3 Action Futility below) further leads to "consent desensitization" where users blindly consent to risks to which they would not normally not consent to, but for their continuous exposure to consent information overload [24]. When it comes to consent in the digital world convenience is a huge driving factor [26], especially when privacy and consent notifications are considered a nuisance or mistaken for the only gateway to gain access to a digital platform or service [27].

Contrary to legal safeguards that apply to services offered to the general public, such as aviation, the internet is run by a transnational private regime of multinational bigtech companies, next to nation states who tries to govern the safety of their respective citizens via a state-centric standard constitutional regulatory regime. Thus, to ensure the safety of users using the internet in much the same way as aviation regulations, for example, govern the safety of aircraft for the general public, the borderless and global digital ecosystem of the internet requires a centralized and tailored digital constitution to govern the internet independently from the many political, business, and financial agendas of the current role players [34]. Only when users of the internet can enjoy a broad spectrum of fundamental digital rights in terms of such a digital constitutional may they enjoy protection from general risks and threats specific to the digital environment, which will equally minimizing the risks exposed to when consenting. However, even this idealistic framework of digital fundamental rights may still not excuse data controllers from the ethical requirements of obtaining consent as discussed above.

2.2 Consideration and Comprehension

The language used to communicate consent and privacy notifications is often unclear and confusing to the extent that very few users may really understand the nature, extent and reason of the information or digital services exchange that is set in motion through their digital consent. As seen in the biomedical field, a proper consent process takes a lot of time, effort and energy to allow a patient or participant to gain full understanding on the level required for purposes of ethical consent. In this regard a study by Braddock et al. reported a median consenting time of 16 min for orthopaedic surgery [21], whilst consenting times for clinical trials has been established at around 30–60 min [22]. Consenting in the digital world usually requires a decision to be made in a very short period of time, if not immediately, leaving very little to no time to carefully consider the consequences of any decision. In addition comprehension of vaguely worded or inherently consent notices may be even less among vulnerable populations such as children, users with low digital literacy, users experiencing language and cultural diversity or obstacles, the cognitively impaired, or users under extreme emotional distress [23].

2.3 Action Futility

Solove also argues that even if the above challenges in respect of habituation and information overload, and the lack of time to adequately consider and comprehend information were to be cleared out of the way, that any actions users may attempt to take will still be futile [16]. Information exchange and comprehension that should serve as enablers of users' autonomy through decision making are seemingly impossible in the digital world due to the lack of real control users have over the outcomes of their digital decisions. In the absence of moral or ethical consent, as discussed above, and regardless of the fact that such consent may be deemed legal, Solove considers this so-called "privacy self-management" as meaningless [16], especially considering that the prevailing approach to privacy merely attempts to protect an individual's choice by remaining neutral on the types of policies accepted. Nissenbaum and Barocas echo Solove's opinion to the extent that they also deem informed consent, if it was achievable, to be ineffective against contemporary information harms on the basis that modern data practices are premised on future and unanticipated or unknown uses [9].

It is understandable that online platforms wish to create a smooth, irritation free user experience for their users, but does it also seem that such an experience is not compatible with ethical consent requirements due to the lack of adequate information exchanges that will allow users to exercise well considered and autonomous consent [28]. The more 'seamless' and enjoyable the experience for the user, the less likely they have contended with complex information requiring them to make difficult choices [29]. Subsequently, consent in the digital world may not provide the 'safety self-management' it is advocated to provide [30].

3 Conclusion

The GDPR's aim is to give control back to individuals over their own personal data. However, by merely stating that it will try to achieve this by placing more emphasis on the obligations of the data controller to operate in a *transparent and lawful* manner will definitely ensure legal consent, but not necessarily ethical consent and true individual control and autonomy as intended. Transparency is often implemented in practice by simply providing online users with more and more information, leading to information overload, habituation and desensitization, achieving the complete opposite of the intended goal of transparency. However, these actions, and the subsequent consent obtained as a result thereof, will be considered to be completely lawful as it ticks the legal requirement of transparency. The GDPR is mute on any obligations of the data controller to ensure that data subjects fully comprehend the information provided to them, as is the case with biomedical treatment and clinical research consent.

As discussed, ethical consent has the ability to transform the moral landscape between two parties. However, if one of the parties is not even aware of this transformation because his or her consent has been automated or "extracted" via digital consent tools, the ethically envisioned transformation could not take place and will the consent also be unethical. The overor hyper-automation of consent may thus undermine the control that an individual may gain over his or her information which is a legally protected right in terms of the GDPR and an essential element of any theory of consent.

Digital consent tools should thus aim to empower online users to better understand notifications and what platforms intend to do with any personal information they collect. Digital consent models should also aim to be interactive and able to engage with users to answer their privacy or information sharing related questions. However, digital consent models still only offer an initial, one-time conclusive consent in an information environment that demands continual consent. Ongoing engagement with users should be sought via a more cooperative model.

Authors such as Pöhls and Rakotondravony proposed some technical solutions to enable users to control data collection in a smart device connected environment by recommending the use of physical kill switches, physical status indicators, and mixes of switches and status indicators to "legally opt out of data collection dynamically due to changes in the privacy-invasiveness tolerability of users due to changes in their situations" [35]. This concept of physically "refusing" consent and frustrating the subsequent collection of data through visual, haptical, or audio feedback depends on the underlying assumption that "the user understands what level of data collection means what level of privacy-invasiveness" [35]. This assumption so clearly illustrates the misconception of ethical consent in many of the technical solutions proposed for consent on the internet, and the misunderstanding of legal versus ethical consent. The switches proposed by Pöhls and Rakotondravony may be an effective technical means of physically preventing further data collection through sensors connected to the Internet of Things (IoT), but if these methods are offered as solutions to "Dynamic Consent: Physical Switches and Feedback to Adjust Consent to IoT Data Collection" as the title of their paper suggests, it clearly falls short of the ethical requirement of ensuring that users *understand* what they are consenting to, based on their own assumption, quoted above.

Coming back to the Nuremberg Code and its emphasis on the voluntary nature of consent and its prohibition of "any element of *force*, fraud, deceit, duress, *overreaching*," "constraint" or "coercion", it is well worth mentioning that many of the digital consent models seem to have embedded at least elements of *force* and *over-reaching* into their models, as discussed in Sects. 2.1, 2.2 and 2.3 above. This code further clearly places the obligation to ensure that participants (users in the case of digital environments) are provided with "sufficient knowledge *and comprehension about the elements*" of the information provided on the "individual [digital platforms] who initiates, directs or engages in the experiment [digital consent transaction]". Ethically speaking the redesign of digital consent models is the task of digital platform operators and data controllers, with other words the parties with the knowledge about the functionalities of the platform, how and what data it collects from users and more importantly how this data will be used. It is thus the responsibility of the digital platform party to rebalance the current informational inequities by redesigning digital consent models.

Interestingly enough, regardless of the existence of the Nurmeberg Code, doctors still circumvented this code, which ignorance gave birth to The Belmont Report that aimed to further establish and extend existing informed consent practices in an effort to protect individual control and decisional power over his or her *body*. In this regard the Belmont Report still protects the control and decisional power over people's bodies, but in the digital age people need control and decisional power over their personal information as well, which may not be explicitly covered by The Belmont Repot or Nuremberg Code. This begs the question whether another uprising must not give birth to a new version or similar instrument as The Belmont Report that can again explicitly and further establishes and extend existing consent principles to now also provide for the ethical application of consent in the digital age and environment.

Solving the issue of ethical consent in the digital world is clearly complex and will not only be found in proposed technical solutions. This requires a much more fundamental and conceptual approach. In this regard I think the following should be considered to take us a few steps closer to finding a workable solution:

1. **Digital constitution** A digital constitution can return political concerns and per-spectives, informed by economic and technical realities of the internet, back into the governance of the internet, and ground the political struggle over the internet explicitly in the fundamental rights of individuals;
2. **Collective consent** Considering that most, if not all, users of the internet can be ren-dered vulnerable at some point in time by manipulating techniques or technologies, a protecting authority that safeguards the interest of users as a collective should be considered;
3. **Digital literacy** To bridge the digital divide, more effort must be put into digital literacy to empower users through insight knowledge of how the internet works, its role players, and the value and use of users' data.

Acknowledgements. This work has been funded by the Luxembourg National Research Fund (FNR)—IS/14717072 'Deceptive Patterns Online (**Decepticon**); the European Union's Hori-zon 2020 Innovative Training Networks, Legality Attentive Data Scientists (**LEADS**) under Grant Agreement ID 956562; and **REMEDIS** Project INTER/FNRS/21/16554939/REMEDIS (Regulatory solutions to MitigatE DISinformation).

References

1. Vollmann, J., Winau, R.: Informed consent in human experimentation before the Nuremberg code. BMJ. **313**(7070), 1445–1449 (1996). https://doi.org/10.1136/bmj.313.7070.1445
2. The Nuremberg Code (1996). The nuremberg code (1947)
3. Brandt, A.M.: Racism and Research: the case of the Tuskegee syphilis study. Hastings Cent. Rep. **8**(6), 21–29 (1978). https://doi-org.proxy.bnl.lu/3561468
4. The Belmont Report. Ethical Principles and Guidelines for the Protection of Human Subjects of Research. The National Commission for the Protection of Human Subjects of Biomedical and Behavioral Research (1979)
5. The General Data Protection Regulation (EU) 2016/679 on the protection of natural persons with regard to the processing of personal data and on the free movement of such data, and repealing Directive 95/46/EC. Recital 7
6. Clifford, D., Graef, I., Valcke, P.: Pre-formulated declarations of data subject consent—citizen consumer empowerment and the alignment of data, consumer and competition law protections. German Law J. **20**, 682 (2019)
7. The General Data Protection Regulation (EU) 2016/679 on the protection of natural persons with regard to the processing of personal data and on the free movement of such data, and repealing Directive 95/46/EC. Articles 13 and 14
8. Beauchamp, T.L., Childress, J.F.: Principles of Biomedical Ethics, 7th edn. Oxford University Press, Oxford (2013)
9. Barocas, S., Nissenbaum, H.: Big data's end run around procedural privacy protections. Commun. ACM **57**(11), 31–33 (2014)
10. The General Data Protection Regulation (EU) 2016/679 on the protection of natural persons with regard to the processing of personal data and on the free movement of such data, and repealing Directive 95/46/EC. Article 12
11. Jones, M.L., Kaufman, E., Edenberg, E.: AI and the ethics of automating consent. IEEE Secur. Priv. Mag. **16**(3), 64–72 (2018). https://doi.org/10.1109/MSP.2018.2701155
12. Miller, F.G., Wertheimer, A. (eds.): The Ethics of Consent: Theory and Practice. Oxford University Press, Oxford (2010)
13. Yao-Huai, L.: Privacy and data privacy issues in contemporary China. Ethics Inf. Technol. **7**, 7–15 (2005). https://doi-org.proxy.bnl.lu/10.1007/s10676-005-0456-y
14. Warf, B.: Geographies of global Internet censorship. Geo J. **76**, 1–23 (2011). https://doi.org/10.1007/s10708-010-9393-3
15. Fitzpatrick, E.F.M., Martiniuk, A.L.C., D'Antoine, H., Oscar, J., Carter, M., Elliott, E.J.: Seeking consent for research with indigenous communities: a systematic review. BMC Med. Ethics **17**, 65 (2016)
16. Solove, D.J.: Privacy self-management and the consent dilemma. Harv. L. Rev. **126**(7), 1880–1903 (2013)
17. Bravo-Lillo, C., Cranor, L., Komanduri, S., Schechter, S., Sleeper, M.: Harder to ignore? Revisiting po p-up fatigue and approaches to prevent it. In: 10th Symposium on Usable Privacy and Security, pp.105–111 (2014). https://www.usenix.org/conference/soups2014/proceedings/presentation/bravo-lillo. Accessed 28 Feb 2022
18. Lindegren, D., Karegar, F., Kane, B., Pettersson, J.S.: An evaluation of three designs to engage users when providing their consent on smartphones. Behav. Inf. Technol. **40**(4), 398–414 (2019). https://doi.org/10.1080/0144929X.2019.1697898
19. Varnhagen, C.K., et al.: How informed is online informed consent? Ethics Behav. **15**(1), 37–48 (2005). https://doi.org/10.1207/s15327019eb1501_3
20. Frauenstein, E.D., Flowerday, S.V.: Social network phishing: becoming habituated to clicks and ignorant to threats? IEEE (2016). https://ieeexplore-ieee-org.proxy.bnl.lu/stamp/stamp.jsp?tp=&arnumber=7802935. Accessed 28 Feb 2022

21. Braddock, C., Hudak, P.L., Feldman, J.J.B.S., Frankel, R.M., Levinson, W.: Surgery is certainly one good option: quality and time-efficiency of informed decision making in surgery. J. Bone Joint Surg. Am. **90**, 1830–1838 (2008)
22. McNair, L., Costello, A., Crowder, C.: Electronic informed consent: A new industry standard (2014). http://docplayer.net/12830094-Electronic-informed-consent-a-new-industry-sta ndard.html. Accessed 28 Feb 2022
23. Bester, J., Cole, C.M., Kodish, E.: The limits of informed consent for an overwhelmed patient: clinicians' role in protecting patients and preventing overwhelm. AMA J. Ethics **6**(18), 869–886 (2016)
24. Schermer, B.W., Custers, B., van der Hof, S.: The crisis of consent: how stronger legal protection may lead to weaker consent in data protection. Ethics Inf. Technol. **16**(2), 171–182 (2014)
25. Obar, J., Oeldorf-Hirsch, A.: The biggest lie on the Internet: Ignoring the privacy policies and terms of service policies of social networking services. Commun. Soc. **23**, 128–147 (2020)
26. Mulder, T.: Privacy policies, cross-border health data and the GDPR. Inf. Commun. Technol. Law **28**(3), 261 (2019)
27. Brandimarte, L., Acquisti, A., Loewenstein, G.: Misplaced confidences: privacy and the control paradox personality. Science **4**(3), 340 (2013)
28. Thaler, R.H., Sunstein, C.R.: Nudge: Improving Decisions About Health, Wealth & Happiness. Penguin (2008)
29. Watcher, S.: The GDPR and the internet of things: a three-step transparency model. Law Innov. Technol. **10**(2), 266 (2018)
30. Peppet, S.R.: Regulating the Internet of things: first steps toward managing discrimination. Priv. Secur. Consent **93**(1), 85 (2014)
31. Macrae, D.J.: The council for international organizations and medical sciences (CIOMS) guidelines on ethics of clinical trials. Proc. Am. Thorac. Soc. **4**, 176–179 (2007)
32. The General Data Protection Regulation (EU) 2016/679 on the protection of natural persons with regard to the processing of personal data and on the free movement of such data, and repealing Directive 95/46/EC. Regulation 58
33. The General Data Protection Regulation (EU) 2016/679 on the protection of natural persons with regard to the processing of personal data and on the free movement of such data, and repealing Directive 95/46/EC. Article 7 and Recital 32
34. Botes, W.M.: Brain Computer Interfaces and Human Rights: Brave new rights for a brave new world. In: FAccT 2022, 21–24 June 2022, Seoul, Republic of Korea. Association for Computing Machinery (ACM). https://doi.org/10.1145/3531146.3533176
35. Pöhls, H., Rakotondravony, N.: Dynamic consent: physical switches and feedback to adjust consent to IoT data collection. In: Streitz, N., Konomi, S. (eds.) HCII 2020. LNCS, vol. 12203, pp. 322–335. Springer, Cham (2020). https://doi.org/10.1007/978-3-030-50344-4_23

HoneyGAN: Creating Indistinguishable Honeywords with Improved Generative Adversarial Networks

Fangyi Yu(✉) and Miguel Vargas Martin(✉)

Ontario Tech University, Oshawa, ON L1G 0C5, Canada
{fangyi.yu,miguel.martin}@ontariotechu.ca

Abstract. Honeywords are fictitious passwords inserted into databases in order to identify password breaches. Producing honeywords that are difficult to distinguish from actual passwords automatically is a sophisticated task. We propose a honeyword generation technique (HGT) called HoneyGAN and an evaluation metric based on representation learning for measuring the indistinguishability of fake passwords, together with a novel attack model for evaluating the efficiency of HGTs. We compare HoneyGAN to state-of-the-art HGTs proposed in the literature using both evaluation metrics and a human study. Our findings indicate that HoneyGAN creates genuine-looking honeywords, leading to a low success rate for knowledgeable attackers in identifying them. We also demonstrate that our attack model is more capable of finding real passwords among sets of honeywords compared to previous works.

Keywords: authentication · machine learning · honeywords

1 Introduction

Current password-based authentication systems store sensitive password files that make them ideal targets for attackers because if successfully obtained and cracked, an adversary may impersonate registered users undetectable [11]. To effectively detect password leaks, Juels and Rivest [4] suggest that a website could store decoy passwords, called *honeywords*, alongside real passwords in its credential database, so that even if an attacker steals and reverts the password file containing the users' hashed passwords, they must still choose a real password from a set of k distinct *sweetwords*, where a real password and its associated honeywords are referred to as sweetwords. The attacker's use of a honeyword could cause the website to become aware of the breach. Notably, honeywords are only beneficial if they are difficult to distinguish from real-world passwords; otherwise, a knowledgeable attacker may be able to recognize them and compromise their security. Thus, when implementing this security feature into current authentication systems, the honeyword generating technique is critical.

The following are the paper's key contributions:

© The Author(s), under exclusive license to Springer Nature Switzerland AG 2023
G. Lenzini and W. Meng (Eds.): STM 2022, LNCS 13867, pp. 189–198, 2023.
https://doi.org/10.1007/978-3-031-29504-1_11

- We propose HoneyGAN, an HGT leveraging a password guessing model called GNPassGAN [14]. HoneyGAN can create passwords that seem legitimate and could be used in a honeyword system to deceive attackers.
- We introduce two evaluation metrics for determining the indistinguishability of honeywords and compare the honeywords generated by our technique HoneyGAN to those generated by other two state-of-the-art HGTs in the literature, and so could reliably infer our framework's true resistance to sophisticated discriminating attackers.
- We conducted a human study via Amazon Mechanical Turk to test the difficulty of finding the real passwords in sets of honeywords created by our HGT and other two state-of-the-art HGTs. Our findings are consistent with the result of using the two evaluation metrics we proposed. To the best of our knowledge, we are the first to conduct a research ethics-approved human participant study related to honeywords.
- To encourage more research on this area and to improve reproducibility, we have made the source code[1] for HoneyGAN publicly available.

The remainder of the paper is structured as follows: Sect. 2 introduces HoneyGAN, and two other HGTs for comparison. Section 3 is the HGTs evaluation. Section 4 is the user study and Sect. 5 discusses the limitations and future prospects of our study.

2 Honeyword Generation Techniques

2.1 HoneyGAN

GNPassGAN. Our HGT is inspired by a password guessing model GNPass-GAN [14]. GNPassGAN is a GAN-based model that consists of a discriminator (D) and a generator (G) that are both constructed using deep learning neural networks. G takes as inputs noise or random features, learns the probability of the input's features, and creates data that follow the distribution of the input data. While D gets both real passwords and samples generated by G and makes every attempt to distinguish the two by calculating the conditional probability of a sample being false (or real) given a set of inputs (or features). This cat-and-mouse game forces D to extract vital information from the training data, and each iteration brings G's output closer to the distribution of real passwords, improving the possibility of matching the passwords of real-world users. GNPass-GAN also incorporates gradient normalization to boost its guessing capability.

GNPassGAN is adept at generating realistic passwords, with 12.65% of passwords created by GNPassGAN being confirmed to exist in real-world password breaches (the *Rockyou* test set) [14], and the generated passwords that do not match the test set are plausible candidates for human-generated passwords. Because the primary challenge of honeyword creation is to develop indistinguishable decoy passwords that attackers cannot discern apart from genuine ones, theoretically, we reckon GNPassGAN can be employed for this purpose and demonstrate it quantitatively in Sect. 3 and 4 via experiments. The main difference

---

[1] https://github.com/fangyiyu/HoneyGAN.

between GNPassGAN and HoneyGAN is that GNPassGAN is used as a password guessing tool in our work, and HoneyGAN is a HGT that utilizes only the generator of GNPassGAN to generate honeyword candidates, and further select the passwords that are most similar to the real password as honeywords.

Text Similarity. Similarity between two strings is crucial in HGT since it demonstrates the indistinguishability of a false password from a genuine one, and is employed in both the honeyword creation and assessment processes. Typically, in natural language processing tasks, the distance/similarity of two strings is determined as follows: the strings are converted to vectors using word embedding techniques, and then the cosine similarity of the two vectors is calculated as the distance. Here, the strings might be composed of letters, symbols, or numbers, similar to how passwords are composed. Popular word embedding methods include Word2vec [6], FastText [1], and $TF - IDF$. While these techniques take into account the semantic and syntactic meanings of a word/text, in our case, the majority of passwords lack such meanings; hence, we choose the simplest but still effective method of vectorization known as bag of words (BoW).

In BoW, the core premise is that documents are similar if they contain comparable information. We examine the histogram of the characters included inside the strings, that is, each character count is considered as a feature. To be more precise, we first count the unique characters and their occurrences in the two strings being compared, then create a vector for each string with a length equal to the number of unique characters the strings contain, assign the vector's value in the associated index to the character's occurrences in each string, and finally compute the cosine similarity of the two vectors by definition. Please note that we do not consider the semantic connotations of passwords in this work.

Generate Honeywords with GNPassGAN. The following procedure demonstrates how we generate honeywords using GNPassGAN. (1) GNPassGAN first needs to be trained on a password corpus, and we train GNPassGAN for 200,000 iterations to get a thorough grasp of the construction pattern of passwords in the training dataset. (2) We use the GNPassGAN generator to produce a file F containing 50,000 fake passwords as honeyword candidates. Notably, F must be stored separately from the authentication system in a secure place. (3) We compare each user's true password to all fake passwords in F and calculate text similarity scores. Here, we convert each password to a vector using BoW and compute the cosine similarity of two passwords. (4) Finally, we assign honeywords for a real password to the $k - 1$ most similar fake passwords in F.

2.2   Baseline Models

We utilize two models as comparisons in this work: chaffing-by-fasttext proposed by Dioysiou et al. [2] and chaffing-by-tweaking proposed by Juels and Rivest [4]. We will use the term chaffing-by-fasttext and *fasttext* interchangeably, as well as chaffing-by-tweaking and *tweaking*.

Chaffing-by-fasttext first trains the *fasttext* model with a real password corpus, then *fasttext* generates vector representations of each password in the corpus. After training is complete, the trained model can be queried by providing a real password as input and receiving a multi-dimensional vector representing the provided password's word embedding as a response. Finally, the top $k - 1$ closest neighbours according to cosine similarity are assigned as honeywords for each password.

Notably, the technique's primary weakness is that the produced honeywords are all genuine passwords in the *fasttext* training dataset, which means that if an attacker has access to the training dataset, the honeywords will be readily discovered. Additionally, the size of the training data has a significant impact on the quality of the honeywords created.

Chaffing-by-tweaking is an approach that mainly relies on random letter, digit, and symbol substitution.

Honeyword examples generated by the three HGTs can be found in Table 1.

Table 1. Honeyword samples generated by the three HGTs compared in the paper (HoneyGAN, *fasttext* and *tweaking*). Our password guessing model GNPassGAN and the *fasttext* model have been trained on a subset of the *Rockyou* dataset.

Real Passwords	deshaun96	dafnny_24	Shauni16!
HoneyGAN	masdane69	andey124	nahuas11
	sandesh89	badhyn24	hunhzan1
	naueds09	maydona242	hanilin1
fasttext	boedha21	snuffy22	muchluv!
	cutechica1	Dushido07	cliffordx
	felli1330	Dampire2	10.04.88
tweaking	DeShauN37	dafnny=96	Shauni53+
	deshaun87	dafNnY44	SHaunI73$
	DesHaun56	dAfnny+47	SHaUnI73$

3 Evaluation

3.1 Datasets

We analyze HoneyGAN's performance and compare it to the other two HGTs using 13 datasets containing real-world passwords. Our password datasets include over 828 million plain-text passwords and are derived from 13 different online providers (can be found in Table 2). We analyze these datasets and choose only passwords with a length of more than 8 characters, and we randomly choose 10,000 authentic passwords from each disclosed dataset to facilitate the assessment of the HGTs without sacrificing generality.

3.2 Internal Similarity Between Honeywords and Real Passwords

The primary goal of HGTs is to create indistinguishable fake passwords; that is, the honeywords and their corresponding actual passwords are too close to be differentiated. Consider passwords to be texts; we can determine the similarity of two passwords by comparing their text similarities. The greater the similarity score, the more similar the two passwords are, and the more difficult it is to distinguish them. We use the BoW metric to determine the similarity of two words without considering the semantic and syntactic meanings.

However, this metric is based on the assumption that an attacker attempts to differentiate real passwords using no resources. Indeed, they may have accessed a large number of previously compromised password files from data breaches. Because 40% of users reuse their passwords [7], more sophisticated attackers would assault the sweetwords using these accessible passwords. As a result, we develop an attack model as described in Sect. 3.3 and assess the resilience of the HGTs based on the aforementioned assumption of attackers. The performance of an HGT is then determined by combining these two evaluation metrics.

3.3   Attack Model: Normalized Top-SW

Our attack model, termed Normalized Top-SW is inspired by Wang et al.'s work "Normalized Top-PW" [11], and operates as follows: 1) Consider a genuine password dataset (attack) obtained from a data breach, and the sweetword file (target). The attacker employs the BoW to vectorize all passwords and sweetwords. 2) The attacker calculates the cosine similarity between each sweetword in the target file and all genuine passwords in the attack dataset, and then assigns the maximum similarity score to the sweetword denoting the highest likelihood of it being a true password. 3) The attacker tries the sweetwords of each user in decreasing order of their scores. If the guessed sweetword is a valid password for the associated user, then delete this user from the dataset; otherwise, set the similarity of the guessed sweetword to 0 to prevent it from being tried again.

In our experiment, we determine the efficiency of HGTs by computing the attacker's success rate under various attempts T. More precisely, we count the number of user accounts that are successfully cracked under varying T assignments and divided by the total number of users to get the attack success rate. We place all genuine passwords in the first column of the sweetword file for the simplicity of evaluation; in practice, operators should shuffle the order of sweetwords and securely keep the index of the real passwords.

3.4   Results

As recommended in [4], we assign $k = 19$ honeywords to each user and calculate the internal similarity score for each sweetword file generated by the three HGTs. Assume we are the *Rockyou* system operator and train our GNPassGAN and *fasttext* on our own dataset (*Rockyou*) to create honeywords for our users. We then attack the produced sweetword file using all other datasets in Table 2. For each user, the attacker has $T = 20$ attempts.

Average Internal Similarity. As a result, the internal similarity score for honeywords created by chaffing-by-GNPassGAN (HoneyGAN) is 0.8193, whereas chaffing-by-fasttext is 0.2620, and chaffing-by-tweaking is 0.6270. These numbers

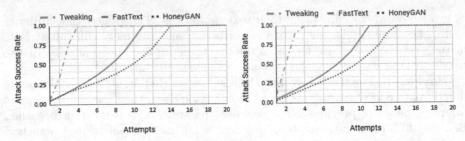

(a) Attack Success Rate using all datasets (b) Attack Success Rate using the *zynga* except for *zynga*.   dataset.

Fig. 1. The Attack Success Rate by using the datasets in Table 2 (except for *Rockyou* as it is the target file) to attack the sweetword file generated by the three HGTs under the Normalized Top-SW attack. A line closer to the y-axis means the HGT is more vulnerable to attacks. As a result, honeywords generated by chaffing-by-tweaking are the easiest to attack, and by HoneyGAN are the hardest.

indicate that the honeywords created by HoneyGAN have the shortest average distance to their corresponding genuine passwords, implying that they are more similar to their true passwords and hence more difficult to differentiate.

Attack Success Rate (ASR). As illustrated in Fig. 1, under our Normalized Top-SW attack, when all datasets except *Rockyou* (exclude it since it is the target) are used as the attack dataset, we see the same pattern: we are able to crack all users' accounts in 4 attempts under the chaffing-by-tweaking condition, in 11 attempts under the chaffing-by-fasttext condition, and in 14 attempts under the HoneyGAN condition. Furthermore, 13 attempts are sufficient for the *zynga* dataset under the HoneyGAN condition. As a result, honeywords formed by *tweaking* are the simplest to discern, while those generated using HoneyGAN are the most difficult.

We show the average attack success rate ($AASR$) in Table 2, where $AASR = \frac{1}{20}\sum_{i=1}^{20} ASR^{(i)}$. As can be seen in the table, an attacker could achieve a success rate of around 60% when honeywords are created using HoneyGAN and 68% when honeywords are generated using *fasttext* when given 20 attempts per user, and it is statistically significant ($p = 3.09 * 10^{-12}$ for a one-tale t-test) that the attack success rate is lower when attacking honeywords generated by HoneyGAN than *fasttext*. Honeywords generated by *tweaking* is the most vulnerable with more than 90% attack success rate. Furthermore, HoneyGAN can produce better undetectable honeywords than *fasttext* and *tweaking* regardless of which dataset is used as the resource for attacking.

HoneyGAN outperforms *fastext* and *tweaking* in terms of both average internal similarity and attack success rate, indicating that HoneyGAN-generated honeywords are more similar to real passwords, therefore deceiving attackers and reducing their attack success rate, and alerting honeycheck towards the password breach.

Table 2. The Average Attack Success Rate on the three HGTs when various attack datasets with *Rockyou* as the target dataset are used. A number in bold indicates that the relevant HGT performs the best.

Dataset	Tweak	FastText	HoneyGAN
have-i-been-pwned-v2	0.9149	0.6863	**0.5923**
linkedin	0.9092	0.6863	**0.5943**
myspace	0.9279	0.6857	**0.6090**
youku	0.9072	0.6858	**0.6090**
zynga	0.9300	0.6907	**0.6213**
adultfriendfinder	0.9230	0.6902	**0.6006**
dubsmash	0.9229	0.6886	**0.6138**
last.fm (2016)	0.9226	0.6854	**0.5880**
chegg	0.9123	0.6888	**0.6032**
dropbox	0.9257	0.6928	**0.6096**
yahoo	0.9188	0.6881	**0.5868**
phpbb	0.9260	0.6855	**0.5972**

4   User Study

4.1   Study Design

We want to validate the hypothesis that individuals need more attempts to correctly find the real password when honeywords are generated by HoneyGAN than *tweaking* and *fasttext*.

We conducted a within-subjects experiment with 300 participants where each person performed all three HGTs. In our experiment, we have one independent variable: HGT type; three conditions: HoneyGAN, *tweaking* and *fasttext*; and one dependent variable: the number of attempts required to find the real password. Our study was approved by the Research Ethics Board at our institution.

Similar to previous security-related studies [3,5,8,10], we recruited participants through Amazon Mechanical Turk (AMT), where we embedded a survey designed on an online survey platform called Qualtrics. Qualified respondents were encouraged to complete our survey. We imposed three requirements on participants: (1) To avoid misunderstandings about our instructions, we need participants to be proficient in English; hence, we required participants exclusively from English-speaking countries including Canada, the US, the UK, and Australia. (2) Participants should have general knowledge as to what secure passwords look like, and we would expect that normally people savvy in information technology have such knowledge. So we only recruited those who self-identify as having a job related to information technology. (3) Additionally, we aim to include only individuals who accomplish high task quality on AMT, as measured by two AMT scores: the total number of approved Human Intelligence Tasks (HITs) and the percentage of approved HITs. We selected individuals who have 1,000 or more approved HITs and a 90% or greater approval rate for HITs.

Participants were required to answer 18 rank-order questions, which match 6 sets of honeyword samples produced from each of the three HGTs. Each question has 19 honeywords and 1 real password. The order of the 20 sweetwords is randomized. The participants were asked to sort the 20 sweetwords in each question according to their level of confidence that the sweetword is a real password. We compensated each participant with CAD$5.00 for completing the experiment, and the compensation was prorated using the Ontario minimum wage at the time of the study.

4.2 Results

Our analysis is based on the responses to our survey that each participant provided. We want to determine if there is a significant difference in the average number of attempts required for users to properly guess the real password in the HoneyGAN condition compared with the other two conditions.

Among all 300 responses, 7 responses were detected as robots by Qualtrics, and we deleted these suspicious responses. The remaining 293 responses took between 47 s and 211 min to complete. To ensure validity, we removed 13 of the 293 replies from participants who finished the exam in less than 3 min, as it is possible that they were not concentrating. Additionally, we eliminated outliers with completion time longer than 39 min and 30 s (boxplot maximum), leaving us with 272 responses to analyze.

The average completion time for the remaining 272 surveys was 14 min with 58 s, with a standard deviation of 7.86 min. This would suggest that the remaining participants were diligent in their responses.

We concatenated the responses for each HGT and got a dataset containing three columns (the three HGTs), and 1632 (6×272) rows, where each value represents the attempts needed to find the real password in one of the questions in the corresponding HGT. Since our experiment is a within-group study with non-uniform data, we used two-factor ANOVA without replication to examine the effect that the HGTs have on attempts needed to find the real password. The results indicated that the type of HGT resulted in statistically significant differences in the number of attempts required to find the real password ($F(2, 3262) = 448.276$, $p \leq 0.001$). We also ran two paired-samples t-tests to examine if there are significant differences between attempts required to find the real password for HoneyGAN vs *tweaking*, and HoneyGAN vs *fasttext*. As a result of comparing HoneyGAN and *fasttext*, the mean number of attempts required to find the real password is 12.479 in the HoneyGAN condition, meaning that participants require approximately 13 attempts to find the real password when HoneyGAN generates the honeywords, compared to 6.734 when *fasttext* generates the honeywords. And the result is statistically significant ($t(1631) = 29.767$, $p \leq 0.001$). A similar result can be found in the comparison of HoneyGAN vs *tweaking*: HoneyGAN requires 12.479 attempts while *tweaking* requires 8.89 ($t(1631)=16.948$, $p \leq 0.001$).

5   Discussion

In this section, we highlight the limitations and future work of our study.

Semantics in Passwords. One limitation of our study is that we did not consider the semantic meanings of passwords. This is flawed when authentication systems incorporate passphrases to assist users in memorization [3,9]. A passphrase, as opposed to a password, is typically a 4-to-10-word phrase, sentence, or statement having semantic and grammatical connotations.

Targeted Attacks. For targeted attacks, attackers exploit users' Personal Identifiable Information (PII) to guess passwords, which increases the likelihood of users' accounts being compromised. This is a critical problem because numerous PII and passwords become widely accessible as a result of ongoing data breaches, and people are used to create easy-to-remember passwords using their names, birthdays, and their variants [12]. Once an attacker obtains users' PII, and if only one sweetword in a user's sweetword list contains the user's PII, it is highly likely that this sweetword is the real password and others are fake.

To the best of our knowledge, Wang et al. [13] are the only ones that discuss how to generate honeywords that are resistant to targeted attacks. We are currently investigating how to generate honeywords for the same purpose with Natural Language Processing techniques.

6   Conclusions

In this paper, we propose HoneyGAN, an HGT built on top of GNPassGAN that generates high-quality honeywords capable of luring attackers and detecting password breaches. HoneyGAN can be easily integrated into any current password-based authentication system. Additionally, we present internal text similarity to assess the quality of honeywords and Normalized Top-SW, a honeyword attack model that mimics the real-world attack situation and avoids any ambiguity. We compare HoneyGAN's performance to two state-of-the-art HGTs using these two metrics, as well as a human study and discovered that Honey-GAN is capable of creating more hard-to-find honeywords and decreasing the success rate of sophisticated attackers. Furthermore, we demonstrated that our attack model Normalized Top-SW is more effective than Normalized Top-PW [11] in discovering real passwords.

Acknowledgement. The authors acknowledge the support of the Natural Sciences and Engineering Research Council of Canada (NSERC), funding reference number RGPIN-2018-05919.

References

1. Bojanowski, P., Grave, E., Joulin, A., Mikolov, T.: Enriching word vectors with subword information. Trans. Assoc. Computat. Linguist. 5, 135–146 (2017)

2. Dionysiou, A., Vassiliades, V., Athanasopoulos, E.: HoneyGen: generating honeywords using representation learning. In: Proceedings of the 2021 ACM Asia Conference on Computer and Communications Security, pp. 265–279 (2021)
3. Jagadeesh, N., Martin, M.V.: Alice in passphraseland: assessing the memorability of familiar vocabularies for system-assigned passphrases. arXiv preprint arXiv:2112.03359 (2021)
4. Juels, A., Rivest, R.L.: Honeywords: making password-cracking detectable. In: Proceedings of the 2013 ACM SIGSAC Conference on Computer and Communications Security, pp. 145–160 (2013)
5. Kelley, P.G.: Conducting usable privacy & security studies with Amazon's mechanical turk. In: Symposium on Usable Privacy and Security (SOUPS), Redmond, WA (2010)
6. Mikolov, T., Chen, K., Corrado, G., Dean, J.: Efficient estimation of word representations in vector space. arXiv preprint arXiv:1301.3781 (2013)
7. Pearman, S., et al.: Let's go in for a closer look: observing passwords in their natural habitat. In: Proceedings of the 2017 ACM SIGSAC Conference on Computer and Communications Security, pp. 295–310 (2017)
8. Redmiles, E.M., Kross, S., Mazurek, M.L.: How well do my results generalize? Comparing security and privacy survey results from MTurk, web, and telephone samples. In: 2019 IEEE Symposium on Security and Privacy (S&P), pp. 1326–1343. IEEE (2019)
9. Shay, R., et al.: Correct horse battery staple: exploring the usability of system-assigned passphrases. In: Proceedings of the 8th Symposium on Usable Privacy and Security, pp. 1–20 (2012)
10. Tuncay, G.S., Qian, J., Gunter, C.A.: See no evil: phishing for permissions with false transparency. In: 29th USENIX Security Symposium (USENIX Security 2020), pp. 415–432 (2020)
11. Wang, D., Cheng, H., Wang, P., Yan, J., Huang, X.: A security analysis of honeywords. In: Network and Distributed System Security (NDSS) Symposium 2018, pp. 1–16 (2018)
12. Wang, D., Zhang, Z., Wang, P., Yan, J., Huang, X.: Targeted online password guessing: an underestimated threat. In: Proceedings of the 2016 ACM SIGSAC Conference on Computer and Communications Security, pp. 1242–1254 (2016)
13. Wang, D., Zou, Y., Dong, Q., Song, Y., Huang, X.: How to attack and generate honeywords. In: 2022 IEEE Symposium on Security and Privacy, pp. 489–506. IEEE (2022)
14. Yu, F., Martin, M.V.: GNPassGAN: improved generative adversarial networks for trawling offline password guessing. In: 2022 IEEE European Symposium on Security and Privacy Workshops (EuroS&PW), pp. 10–18 (2022). https://doi.org/10.1109/EuroSPW55150.2022.00009

Author Index

Printed in the United States
by Baker & Taylor Publisher Services